BIOETHICS IN SINGAPORE

The Ethical Microcosm

A 10th Anniversary Publication of the Bioethics Advisory Committee, Singapore

John M Elliott
W Calvin Ho
Sylvia SN Lim
editors

BIOETHICS IN SINGAPORE

The Ethical Microcosm

NEW JERSEY • LONDON • SINGAPORE • BEIJING • SHANGHAI • HONG KONG • TAIPEI • CHENNAI

Published by

World Scientific Publishing Co. Pte. Ltd.
5 Toh Tuck Link, Singapore 596224
USA office: 27 Warren Street, Suite 401-402, Hackensack, NJ 07601
UK office: 57 Shelton Street, Covent Garden, London WC2H 9HE

Library of Congress Cataloging-in-Publication Data
Bioethics in Singapore : the ethical microcosm / editors, John M. Elliott, W. Calvin Ho, Sylvia S.N. Lim.
p. ; cm.
Includes bibliographical references and index.
ISBN-13: 978-981-4327-10-7 (hardcover : alk. paper)
ISBN-10: 981-4327-10-7 (hardcover : alk. paper)
ISBN-13: 978-981-4327-11-4 (pbk. : alk. paper)
ISBN-10: 981-4327-11-5 (pbk. : alk. paper)
1. Bioethics--Singapore. 2. Medical ethics--Singapore. I. Elliott, John, 1945–
II. Ho, W. Calvin. III. Lim, Sylvia S.N.
[DNLM: 1. Bioethical Issues--Singapore. WB 60]
QH332.B51728 2010
174'.957095957--dc22

2010027032

British Library Cataloguing-in-Publication Data
A catalogue record for this book is available from the British Library.

Copyright © 2010 by National University of Singapore

All rights reserved. This book, or parts thereof, may not be reproduced in any form or by any means, electronic or mechanical, including photocopying, recording or any information storage and retrieval system now known or to be invented, without written permission from the Publisher.

For photocopying of material in this volume, please pay a copying fee through the Copyright Clearance Center, Inc., 222 Rosewood Drive, Danvers, MA 01923, USA. In this case permission to photocopy is not required from the publisher.

Typeset by Stallion Press
Email: enquiries@stallionpress.com

Printed in Singapore.

Contents

Foreword

In December 2000, the Singapore Government established the Bioethics Advisory Committee (BAC). The BAC had the wide remit of investigating and recommending on the ethical, legal, and social issues surrounding the field of biomedical sciences research in Singapore. In the subsequent decade, the Committee investigated a number of pressing matters, starting with the sometimes vexed issue of stem cell research, and its recommendations have been the basis for securing the ethical foundations of burgeoning biomedical research in Singapore.

While the immediate work of the BAC was to recommend an ethical governance framework to regulate this development, this could not be done without an in-depth consideration of issues in a wider context. This book is the product of these considerations. It tries to put the work of the BAC into a broader cultural, social, and ethical context. It is not just a celebratory account of ten years of work, nor is it entirely retrospective in focus, nor is it about Singapore alone. Some chapters are relatively local in their context, but others are strongly international. Almost all of the authors have direct experience in the work of the BAC, and their chapters provide an in-depth account of aspects of their expertise that have impacted the BAC and helped to set it in perspective.

It is worth reiterating that the BAC remit refers to "ethical, legal, and social issues". The first three chapters address such issues from three very different perspectives. In the first two chapters, the respective authors have conducted extensive interviews with BAC members (past and present) and with researchers and research administrators. These yield interesting reflections on the role and impact of the BAC. One clear conclusion from these chapters is that as Singapore moves forward the BAC will need to adapt its *modus operandi* to changes in the character of society, and that changes to the consultation process will need to be a part of this. The third chapter is an unusual one in looking at the work of the BAC and how it engages the public from the perspective of working journalists.

These first three chapters all highlight aspects of the BAC that reflect its Asian character, which shows in its style and non-confrontational approach. The fourth chapter offers an explicit philosophical discussion of Confucian ethics, and what it means for the biomedical framework in Singapore. This chapter will be of great interest to ethicists and moral philosophers keen on considering the extent to which ethical principles diverge as a function of culture and history. This and the remaining chapters in the book move increasingly into the details of specific ethical topics.

Chapter 5 by Campbell *et al.* explores the inherent tensions when the roles of researcher and clinician are combined in a single individual. Chapter 6 by Parham and Lo takes the reader to the United States, where Bernard Lo has been at the forefront of regulatory issues in stem cell research ethics. He and his co-author provide a thorough review of the field which they then helpfully link to the development of regulation in Singapore. Chapter 7 by Knoppers *et al.* also deals with genetics and stem cell research, but their focus is on international policy-making. Their review of this topic offers a comprehensive historical account of the major pre- and post-genomic era regulatory developments and they also consider the Canadian and Singapore situations in some detail.

Chapter 8 by Murray and White of the Hastings Center provides an insightful comparison of the ways in which the National Bioethics Advisory Commission (NBAC) in the United States and the BAC addressed the subject of using human biological materials in research. Established in 1996, the NBAC served as the national bioethics advisory body to the Clinton Administration until 2001. Tom Murray was a former Commissioner on the NBAC, hence his views on the subject are especially pertinent. Murray and White further propose a novel four-stage process in thinking about public engagement on bioethical issues.

Chapter 9 by Ho and Bobrow is an interesting venture into the various kinds of human-animal combinations that have been the subject of recent legislation in the UK, and are currently on the horizon in Singapore. Here, as in other chapters, it is evident that an effort at globalising bioethics does not entail downplaying context and culture; it tends rather to highlight them. In the final chapter, Elliott takes a speculative look at possible future developments and how the BAC may react to them. He concludes that a future emphasis may have to be more explicitly on the need to consider the wider impact of research activities, in order to bring them into some kind of a sustainable balance in the long run. The four annexes at the end of the book provide general information on past and present members of the BAC, its International Panel of Experts, its reports, and key scientific and bioethics events in Singapore and abroad.

Singapore is indeed an ethical microcosm. It is also a legal and social microcosm. Running through all the chapters is a palpable sense that while we have arrived at a destination, having achieved a foundation for an agenda of biomedical research, we have also arrived at a departure point. The directions for the future are only now becoming apparent with the emergence of technologies and knowledge that can potentially transform society by transforming people.

Lim Pin

Chairman
Bioethics Advisory Committee
Singapore

Editors' Note and Acknowledgements

The idea of producing a book about the Bioethics Advisory Committee (BAC) and its work almost a decade after its establishment was conceived as early as 2007. The intention was both to review the development of bioethics in Singapore and to consider what comparative value this may have for a wider readership. The ethical governance of rapid scientific change raises issues internationally and not in Singapore alone.

While we have tried to provide a retrospective appraisal of the contribution of the BAC, our main objective is to make a wider substantive contribution to the rapidly emerging field of bioethics. To that end the book includes original contributions from a number of distinguished writers and researchers in ethics, philosophy, law, and the biomedical and social sciences, all of whom have been closely involved in the institutionalisation of bioethics in Singapore. In many chapters there are also contributions from a younger generation of scholars and researchers interested in the field of bioethics.

Although many of the contributors to this book are closely connected to the BAC, this book is not an official account. All opinions and views expressed are personal to the authors and do not necessarily represent the views of the BAC or any other organisation. All links to Internet websites stated in this book have been verified to be correct and active as of 7 June 2010.

We are honoured to have had the opportunity to work with our contributors and thank them for their insights and also their efforts to meet tight deadlines. We are grateful to Professor Lim Pin, Chairman of the BAC, for supporting the effort to frame the work as an original contribution to the field, and for kindly providing the Foreword. In addition, we are grateful to past and present members of the BAC, as well as to many individuals (not all of whom could be explicitly acknowledged here) who have generously supported this project through the sharing of their views, knowledge, and experiences in interviews, discussions, and records. This project would not have been possible without their help. We would also like to thank our editor at World Scientific, Lim Sook Cheng, for her patience with this project, and Siah Rouh Phin of the NUS Office of Legal Affairs, for her support. Our colleagues, Victoria Ang and Charmaine Chan, provided us with crucial editorial assistance, for which we are also grateful. Finally, we are indebted to our families for their support.

The editing of this book has been a sustained collective and collaborative effort, and none of us claims any priority. The editors listed here and on the title page are in alphabetical order by family name.

John M Elliott,
W Calvin Ho &
Sylvia SN Lim

Notes on Contributors

Martin BOBROW, CBE, FRS, FMedSci, is Emeritus Professor of Medical Genetics at Cambridge University and has been an advisor to Singapore's Bioethics Advisory Committee (BAC) since 2001. He was Chair of a UK Academy of Medical Sciences Working Group that was responsible for the report *Inter-species Embryos* (June 2007). He is currently the Chair of another Working Group of the Academy that is examining the use of animals containing human material in scientific research.

Alastair V CAMPBELL, MA, BD, ThD, is the Chen Su Lan Centennial Professor of Medical Ethics and Director of the Centre for Biomedical Ethics in the Yong Loo Lin School of Medicine at the National University of Singapore (NUS). His recent books include *Health as Liberation* (1996), *Medical Ethics* (with DG Jones and G Gillett, 4th Edition, 2005), and *The Body in Bioethics* (2009). He is currently a member of the BAC, and a Board Member of the Singapore Health Sciences Authority and of the National Medical Research Council.

Charmaine KM CHAN, BSc (Hons), graduated from NUS in 2008 with a degree in Life Sciences (specialisation in Biomedical Science). She is currently with the Secretariat of the BAC.

CHANG Ai-Lien, BA, is a Senior Correspondent with the *Straits Times*. She has been a journalist for 16 years, and specialises in reporting on science in Singapore. She is a member of the Genetic Modification Advisory Committee (sub-committee on public awareness) — an expert panel which makes decisions on genetically modified foods, and is part of the Concepts Committee shaping the development of the new Science Centre Singapore. She is also a board member of the Singapore Mental Health Study, a landmark project looking at the state of mental illness in the Republic, and an advisory board member of the NUS's science communication programmes.

Jacqueline CHIN, BPhil, DPhil, is Assistant Professor at the Centre for Biomedical Ethics, NUS, and jointly appointed to the Department of Philosophy. She is a member of the National Transplant Ethics Panel of Laypersons and the National Medical Ethics Committee Workgroup for Advance Care Planning, and Co-Director of the Singapore Ministry of Health-funded CENTRES project for research and support of the work of clinical ethics committees in public and private sector hospitals.

John ELLIOTT, PhD, is a citizen of Singapore and an Associate Professorial Fellow in the Department of Psychology, NUS, where he has researched and taught since 1986. He is a past member of the BAC, the Singapore National Medical Ethics Committee, and the Panel of Advisors to the Juvenile Court, and is currently with the BAC Secretariat.

W Calvin HO, MS (Econ), LLM, is senior research associate with the Secretariat of the BAC. He is also a doctoral candidate at Cornell University and a visiting Fellow at Cambridge University.

Rosario ISASI, JD, MPH, is a Research Associate at the Centre of Genomics and Policy, Faculty of Medicine, Department of Human Genetics at McGill University. Her research interests intersect public health, ethics, law, and science. She has particular expertise in the area of comparative law and international governance issues surrounding regenerative medicine and stem cell research. She has published in *Science, Human Reproduction, Cell Stem Cell, Stem Cell Research, American Journal of Law and Medicine, Journal of Law, Medicine & Ethics,* amongst other journals.

Emily KIRBY, BSc, is a research assistant at the Centre of Genomics and Policy (McGill University) and is currently completing her Bachelor of Civil Law degree at the University of Montreal.

Bartha Maria KNOPPERS, PhD, is Professor of Law and Bioethics in the Faculty of Medicine (Department of Genetics) and Director of the Centre of Genomics and Policy, McGill University. She is an international expert on genomics and policymaking, Chair of the Ethics Working Party of the International Stem Cell Forum, and Officer of the Order of Canada. She is also an advisor to the BAC.

LIM Pin, MA, MD, FRCP, FRACP, is Chairman of the BAC, University Professor and Professor of Medicine at NUS, and Senior Consultant Endocrinologist at the National University Hospital, Singapore.

Sylvia SN LIM, MBBS, is the Head of Secretariat of the BAC. She was involved in the preparation of all the reports published by the BAC.

Edison T LIU, MD, has been the Executive Director of the Genome Institute of Singapore since 2001, and a member of the BAC between 2003 and 2006. A citizen of the United States, he is a permanent resident of Singapore and has held several key positions both in the US and in Singapore. He was the Scientific Director for the Division of Clinical Sciences at the US National Cancer Institute, and Professor of Medicine, Biochemistry, and Genetics at the University of North Carolina at Chapel Hill. In Singapore, Edison Liu was the previous Executive Director of the Singapore Bio-Bank, formerly known as the Singapore Tissue Network, and holds Professorships in both the Faculty of Medicine at NUS and the College of Sciences at Nanyang Technological University. He is the current Chairman of the Board for the Health Sciences Authority, the health regulatory body for Singapore, and has been the longest serving President of the Human Genome Organisation.

Bernard LO, MD, is a Professor of Medicine and Director of the Program in Medical Ethics at the University of California, San Francisco (UCSF), and is an advisor to the BAC. He is also the Co-Chair of the Scientific and Medical Accountability Standards Working Group of the California Institute for Regenerative Medicine, which develops guidelines for stem cell research in California, and a Council member of the Institute of Medicine, USA.

Thomas H MURRAY, PhD, is President of the Hastings Center. Before that he held the Susan Watson Chair in Bioethics and was Director of the Center for Biomedical Ethics in Case Western University's School of Medicine. Dr Murray is the author of over 250 publications, an advisor to the BAC, and a former Commissioner on the US National Bioethics Advisory Commission from 1996 to 2001.

Anh Tuan NUYEN, PhD, is Associate Professor with the Department of Philosophy in the Faculty of Arts and Social Sciences, NUS. He is also a member of the BAC, the Singapore Transplant Ethics Committee, and the NUS Institutional Review Board.

Lindsay PARHAM, BA, is a research assistant in the Program in Medical Ethics at the University of California, San Francisco, and the Program Coordinator for the Greenwall Faculty Scholars Program in Bioethics. She received her BA *magna cum laude* from Cornell University in Comparative Law and Bioethics through Cornell's College Scholar Program. Her interests include reproductive ethics and justice, health law and policies, and medical and legal anthropology.

Judith TAN, BA (Hons), is a Health Correspondent with the *Straits Times*. A journalist of 21 years, she has vast experience in reporting on medical, environmental, and scientific issues for print, radio, and television. In 2002, she took a two-year break from journalism and worked as director of corporate communications with an international hotel group.

VOO Teck Chuan, MA, is an adjunct research fellow with the Centre for Biomedical Ethics, NUS. He has served as rapporteur for a World Health Organization meeting on research ethics, and is a PhD candidate in the Centre for Social Ethics and Policy, School of Law, University of Manchester, under a Wellcome Trust Studentship.

Ross S WHITE, BA, is a 2009 graduate of Davidson College, where he received his degree in political science and medical humanities. He is currently a research assistant at the Hastings Center in the US.

1

The Coming of Bioethics to Singapore

W Calvin Ho and Sylvia SN Lim

Bioethics as an institutional concern with focus on biomedical research involving or pertaining to human beings took root with the establishment of the Bioethics Advisory Committee (BAC) by the Cabinet in December 2000, to provide the government with advice on ethical, legal, and social issues that might arise from research. It broke fresh ground under the national Biomedical Sciences Initiative, announced by the government in June that year. This initiative meant that institutional bioethics would develop in tandem with various other policy initiatives directed at expanding the narrow base of the national economy.[1] In particular, investment in biomedical sciences, alongside electronics, engineering, and chemicals, was considered critical in broadening Singapore's secondary export-oriented industrialisation as increasing competition from the region[2] necessitated a re-shaping of the nation-state's relationship to the global economy and the domestic allocation of resources among industries and major social groups.[3] However, the investment is not without its share of controversies given the uncertain nature of biomedical research, limited talent pool, and ethical conundra.[4]

---

[1] Normile D. Can Money Turn Singapore Into a Biotech Juggernaut? *Science* **297**, 5586 (2002): 1470–1473.

[2] Lim and Gregory identify competitors for inward investment from established firms abroad and for the limited pool of talent to support the growth of the biotechnology sector as Malaysia, Australia, India, Hong Kong, and China. Lim LPL and Gregory MJ. Singapore's Biomedical Science Sector Development Strategy: Is it Sustainable? *Journal of Commercial Biotechnology* **10**, 4 (2004): 352–362, p. 361.

[3] This notion of development strategy is drawn from Gary Gereffi. See Gereffi G. Paths of Industrialization: An Overview, in *Manufacturing Miracles: Paths of Industrialization in Latin America and East Asia*, eds. G Gereffi and DL Wyman. Princeton, NJ: Princeton University Press, 1990, pp. 3–31, p. 23.

[4] Normile D. An Asian Tiger's Bold Experiment, *Science* **316**, 5821 (2007): 38–41, pp. 39 and 41. See also Cao C. Making Singapore a Research Hub, *Science* **316**, 5830 (2007): 1423–1424.

This chapter provides an account of the institutionalisation of biomedical research ethics in Singapore through the work of the BAC. It draws from interviews that have been conducted with past and present members of the BAC and its sub-committees and working group,[5] as well as from the authors' personal experiences, having worked in the Secretariat of the BAC since 2001. The chapter further examines the incremental manner by which an ethical framework (based on five key principles) has been developed by the BAC. For some, this ethical framework has come to characterise bioethics in Singapore. It concludes with a discussion of the significance of bioethics for the public and for the biomedical research community.

The work of the BAC comes within the genre of public bioethics, as it addresses ethical conundra arising from biomedical research by recommending policies that are intended to apply to all members of society.[6] It is not a body that debates foundational issues, such as the fundamental normative values of biomedical science and technology, nor is it concerned with clinical bioethics, although a number of its recommendations do affect clinical practice. Foundational bioethics and clinical bioethics are addressed by other ethical bodies in Singapore. The former lies within the broad remit of the Centre for Biomedical Ethics, which was established in 2005 in the Yong Loo Lin School of Medicine at the National University of Singapore (NUS). The Centre operates under the direction of Alastair Campbell, who is the Chen Su Lan Centennial Professor of Medical Ethics, a position established in honour of one of the nation's best known philanthropists, Dr Chen Su Lan. The BAC collaborates with the Centre on a variety of educational activities for members of ethics review committees or institutional review boards (IRBs), researchers, and the public. In December 2008, the NUS Centre published the inaugural issue of the journal *Asian Bioethics Review* jointly with the Hastings Center, which is one of the field's premier research bodies in the US. Clinical bioethics, on the other hand, has been a concern of the Ministry of Health (MOH), and its National Medical Ethics Committee (NMEC) in particular. Ong Yong Yau, a former Chair of the NMEC, said that the scope of the NMEC

---

[5]A list of past and present members of the BAC and its sub-committees and working group is set out in Annex A of this publication.

[6]Although the BAC is not focused on calling into question the expectations of scientists concerning particular science and/or technology (such as pharmacogenetics in Adam Hedgecoe's paper), it does moderate these expectations with those of the public. See Evans JH. Between Technocracy and Democratic Legitimation: A Proposed Compromise Position for Common Morality Public Bioethics, *Journal of Medicine and Philosophy* **31** (2006): 213–234; and Hedgecoe A. Bioethics and the Reinforcement of Socio-technical Expectations, *Social Studies of Science* **40**, 2 (2010) 163–186.

is narrower than that of the BAC as it is only concerned with issues that arise from medical practice, such as organ donation, end of life issues, and apportionment of healthcare expenditures.[7] Set up in June 1994, the purpose of the NMEC is to assist the medical profession in addressing ethical issues in medical practice and to ensure a high standard of ethical practice in Singapore. However, he indicated that there should be a close working relationship between the BAC and the NMEC as an increasing number of medical practitioners are involved in research, and biomedical research is finding application in clinical settings at a growing pace. The BAC's report on genetic testing and genetic research was in fact prepared in consultation with the NMEC.

Prior to the establishment of the BAC, formal ethics review of biomedical research was somewhat of a novelty, except in the well established procedures for the regulation of clinical drug trials. The BAC's Deputy Chair, Lee Hin Peng, recalled that he was responsible for setting up an *ad hoc* IRB at NUS when he was the Principal Investigator of a research study with American collaborators in 1992.[8] This project-specific *ad hoc* IRB was established to satisfy a requirement of funding by the US National Cancer Institute, that research carried out in Singapore be reviewed by a local IRB. Subsequently, an IRB was set up by the National University Hospital, where he served as Deputy Chair. But it was not until 2003 that NUS established its own IRB with him as its Chair.

GOOD SCIENCE AND THE INSTITUTIONALISATION OF RESEARCH ETHICS IN SINGAPORE

From around the turn of the 21st century (and perhaps even before that), an important change occurred in the way that 'good science' has come to be understood, not only in scientifically advanced countries, but also internationally. In the context of global health, the observation of Maureen Kelley and her collaborators is especially pertinent:[9]

> *"Good science" now means more than rigorous application of scientific methods toward important scientific discoveries. Good*

[7] Interview with Professor Ong Yong Yau, 29 April 2009.
[8] Interview with Professor Lee Hin Peng, 27April 2009.
[9] Kelley M, Rubens CE and the GAPPS Review Group. Global Report on Preterm Birth and Stillbirth (6 of 7): Ethical Considerations, *BioMed Central Pregnancy and Childbirth* **10**, Suppl 1, S6 (2010): 1–19, p. 1.

> *science has also come to mean a deliberate attempt to direct methodologically rigorous science toward the disease burden of the underserved across borders. With this move, the role for ethics in science is becoming more than an important constraint on scientific practice and unintended consequences of unbridled discovery. Ethics can also inform and shape the research agendas for institutions and stakeholders interested in improving the lives and alleviating suffering...*

Ethics has a profound impact on how biomedical sciences are viewed by society. The legitimacy of biomedical sciences depends not only on methodological rigour, but also on the ethical acceptability of their goals and applications. Edison Liu captures the essence of this in his remark, "Science in the absence of humanity is not only irrelevant but dangerous".[10] He explained that science must be consistent with the fundamental values of society as it can be a destructive force if misapplied. Patrick Tan made a similar point in his observation that 'good science' is also a matter of assessing how far scientific pursuit can be justified within an existing system of norms and values.[11] He considered a goal of science to be pushing against knowledge boundaries,[12] so that like 'good art', 'good science' must not only be unquestionable in terms of its intellectual base or rigour, its results and impact must also extend beyond the present. But where a scientific postulation has profound impact on social norms and values, most scientists tend to be wary about pushing too far.

Patrick Tan went on to observe that apart from broader society, scientists are also embedded within their own communities. Due to this and possibly also to the need for funding, scientists tend to be socially conservative by nature and mavericks are rare. 'Good science' emerges as 'spikes' out of the communal effort of smart people working together on an idea within a suitable milieu. Ethical values are important for scientists within their own community, and in their relationship with broader society. As Marilyn Strathern argues, ethics may be understood as 'personal' responsibilities, rights,

[10] Interview with Professor Edison Liu, 8 July 2009.

[11] Interview with Associate Professor Patrick Tan, 26 June 2009.

[12] Dr Lim Bing expressed a similar view. He felt that biomedical research is about "pushing against boundaries", although scientists in general value life. He himself considers all life to be valuable. Interview with Dr Lim Bing, 13 July 2009.

and liabilities which are drawn from more general social sensibilities embedded in all kinds of human interactions and moralities. Consequently, the ends of ethical conduct ensure that, as means, they meet certain criteria in themselves.[13] Philosopher Nuyen Anh Tuan also observed scientists to be socially more conservative than he thought prior to his experience with the BAC.[14] He was surprised to find that scientists in Singapore tended to avoid socially controversial research even if these controversies lacked ethical or philosophical basis. For instance, scientists have been concerned with the moral status of an embryo even though most philosophers do not regard an embryo as having moral status. Upon reflection, he felt that scientists may have been concerned with public backlash since public opinion was influenced by religious views on the subject. In addition, some scientists as members of society may believe in the sanctity of human personhood as beginning from the point of conception.

Lee Eng Hin, who was Dean of Singapore's only medical school at NUS in 2000, said that the government recognised how intricately biomedical sciences have become intertwined with ethics.[15] Having considered a number of national ethics bodies in the English-speaking world, the BAC was appointed by the government to guide its policies on developing the nation's biomedical research capabilities within an ethical and social normative framework that is acceptable both locally and internationally. He indicated that there was some urgency in fully operationalising the BAC as biomedical research activities had increased exponentially and there was a need to establish clear ethical guidelines for such research especially in the controversial area of human embryonic stem cells. For Singapore's research findings to be accepted internationally it was extremely important for Singapore to have a robust ethical framework. Ethical direction and consistency are further important in sustaining the legitimacy of and commitment to an uncertain long-term venture taken up in the interest of the common good. Tan Chorh Chuan, who was Director of Medical Services (or Chief Medical Officer in some countries) at that time, indicated that the government also recognised that the public must be comfortable with the pace of progress in biomedical sciences, which

[13] Strathern M. Accountability . . . and Ethnography, in *Audit Cultures: Anthropological Studies in Accountability, Ethics and the Academy*, ed. M Strathern. London: Routledge, 2000, pp. 279–304, pp. 292–293.
[14] Interview with Associate Professor Nuyen Anh Tuan, 21 April 2009.
[15] Interview with Professor Lee Eng Hin, 9 April 2009.

would accelerate with the adoption of the Biomedical Sciences Initiative.[16] Hence, the work of the BAC was not limited to advising the government, but also involved promoting public trust. In particular, the BAC had the critical task of facilitating public deliberation of bioethical issues through appropriate framing of such issues and the provision of accessible and factually accurate information which served as a starting point for discussion.

Since the time of its inception, the members of the BAC have been composed of men and women of high standing, capable of providing balanced views and with the expertise to address ethical, legal, or social challenges that biomedical research presents. Members were also chosen with a view to public engagement.[17] Lim Pin, who has been the longest serving Vice-Chancellor (President) of NUS, from 1981 to 2000, was appointed the founding Chair of the BAC. He said that he was at first reluctant to take on this position as he wanted to focus on medicine (he being an endocrinologist by training) after many years of service to the University.[18] However, he appreciated the importance of the government's Biomedical Sciences Initiative and recognised that the success of the initiative would also depend on a sound ethical basis upon which this enterprise was to be built. He described the role of the BAC as one of steering the development of the scientific enterprise based on understanding its temperament and characteristics through close interaction with interested stakeholders but without being directive or overbearing. The BAC does not possess a definite legal identity or statutory powers although it is mandated to formulate policies that, if accepted by the government, profoundly influence the course of biomedical research. This characteristic of the BAC is similar to the national bioethics bodies in a number of countries, such as the US. In 1996, members of the US National Bioethics Advisory Commission were appointed by President Bill Clinton,[19] and such appointments have remained the practice through three different administrations since that time. The situation differs in the UK where, as Alastair Campbell observed,[20] membership positions to serve on ethical set-ups like the Nuffield Council on Bioethics and the Biobank Ethics and Governance Council are advertised

[16]Interview with Professor Tan Chorh Chuan, 4 May 2009.
[17]Interview with Ms Tricia Huang, 22 April 2010.
[18]Interview with Professor Lim Pin, 27 April 2009.
[19]For an evaluation of the work of the US National Bioethics Advisory Commission, see Eiseman E. *The National Bioethics Advisory Commission: Contributing to Public Policy.* Arlington, VA: RAND, 2003.
[20]Interview with Professor Alastair Campbell, 15 April 2009.

in furtherance of the Nolan principles.[21] However, he recognised that this approach may not be suitable for Singapore as the UK has a larger pool of expertise to draw on. In addition, there is a longer tradition of volunteerism in policy-related public service among very qualified people in the UK.[22]

Since 2002, the BAC has provided recommendations to the Steering Committee on Life Sciences, which was constituted by the Cabinet in June 2000 as the Life Sciences Ministerial Committee. The Steering Committee is responsible for fostering the development of biomedical sciences through various policy measures, including the coordination of activities of government ministries such as the Ministry of Trade and Industry, the Ministry of Education, and the Ministry of Health. As we have seen, the establishment of the BAC is a proactive initiative by the government to ensure that biomedical research in Singapore is conducted under standards of ethical governance that are acceptable both locally and internationally.

From the time of its inception, the BAC has constituted four sub-committees and a working group (see Fig. 1) to assist it in examining specific issues in more detail. The Human Stem Cell Research Sub-Committee (HSCRS), the Human Genetics Sub-Committee (HGS), and the Publicity and Education Sub-Committee (PES) were formed in 2001. The HSCRS and the HGS concluded their term in 2006. A working group on Human Embryo and Chimera Research (HECR) was formed in 2006 and the Sub-Committee on Research Involving Human Participants was formed in 2007. Their work is still ongoing, as is the work of the PES. Between 2002 to 2008, six sets of recommendations were published by the BAC, and they relate broadly to the subjects of human embryonic stem cell research and cloning, human tissue, research involving human subjects, and genetics.[23] All the recommendations have been prepared after detailed review of international ethical and regulatory norms, ethical and regulatory practices in key jurisdictions, and after public consultation. In addition, they have also been reviewed by the BAC's International Panel of Experts[24] and endorsed by the Biomedical Sciences International

---

[21] Committee on Standards in Public Life, UK, *The First Seven Reports: A Review of Progress*, September 2001. The seven principles of public life have been set out as selflessness, integrity, objectivity, accountability, openness, honesty and leadership.

[22] Bill Bryson, ed. *Seeing Further: The Story of Sciences and the Royal Society*. London: HarperPress, 2010.

[23] See Annex C of this book for a list of reports published by the BAC between 2002 and 2008.

[24] See Annex B of this book for information on members of the International Panel of Experts.

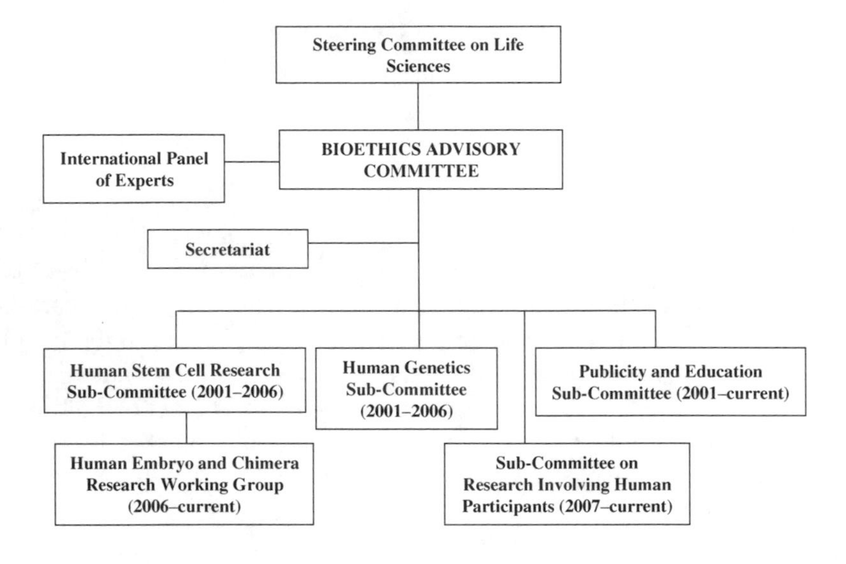

Terms of Reference

The BAC has two main roles:
1. Examine ethical, legal, and social issues arising from research on human biology and behaviour and its applications; and
2. Develop and recommend policies to the Steering Committee on Life Sciences on these issues.

The BAC has three priorities:
1. Protection of the rights and welfare of individuals;
2. Public education and a source of information on bioethical issues; and
3. Identification of broad principles to govern the ethical, legal, and social implications of human biomedical research.

Fig. 1. The Organisation Structure of the BAC and its Terms of Reference.

Advisory Council.[25] These recommendations have since been accepted by the government and all registered medical practitioners who are involved in biomedical research are required to observe the BAC's recommendations under a directive issued by the Director of Medical Services in 2006.[26] The work of the BAC and significant international and local events surrounding its work are listed in the table in Annex D of this book.

---

[25]The Biomedical Sciences International Advisory Council is composed of eminent scientists and visionaries in the field of biomedical sciences. More information is available at: http://www.a-star.edu.sg/AboutASTAR/BiomedicalResearchCouncil/BMSInitiative/BMSIAC/tabid/352/Default.aspx.

[26]Ministry of Health, Singapore, *Directive 1A/2006: BAC Recommendations for Biomedical Research*, 18 January 2006.

RECOMMENDATIONS IN RELATION TO HUMAN STEM CELL RESEARCH AND CLONING

The first set of recommendations on human stem cell research and cloning was prepared by the HSCRS and published by the BAC in June 2002.[27] Richard Magnus (then Senior District Judge), Chair of the HSCRS, explained that the BAC had decided to prioritise its deliberation on stem cell research and cloning because of local research interest in the field. There was also a need to clarify Singapore's position as political contention over the subject intensified.[28] Ng Soon Chye said that local researchers had been involved in IVF-related embryo research for some time. He himself was working with Ariff Bongso on embryo research, and was also interested in therapeutic cloning involving primates.[29] In 2000, the establishment of ES Cell International, which aimed to develop therapies from human embryonic stem cells, drew public attention to Singapore's engagement in this ethically contentious area of embryonic stem cell research. In the year that followed, the then US President George W Bush limited public funding to certain established embryonic stem cell lines and an international treaty to ban cloning was considered by the United Nations.[30] Not surprisingly, many past and present BAC members consider human embryonic stem cell research and cloning to have been the most ethically contentious subject that the BAC considered.

Among the documents considered by the BAC, the Warnock report[31] and a report on the subject by the US National Bioethics Advisory Commission[32] have been important resources. To counter any misperception that Singapore intended to lure stem cell researchers with lax regulatory policies, the BAC was careful not to deviate from the ethical and regulatory policies of major

---

[27] Bioethics Advisory Committee, Singapore, *Ethical, Legal and Social Issues in Human Stem Cell Research, Reproductive and Therapeutic Cloning,* June 2002.

[28] Interview with Mr Richard Magnus, 18 April 2009.

[29] Interview with Professor Ng Soon Chye, 26 May 2009.

[30] For a more detailed analysis of events surrounding the BAC's recommendations on human stem cell research and cloning, see Ho WC. Governing cloning: United Nations' Debates and the Institutional Context of Standards, in *Contested Cells: Global Perspectives on the Stem Cell Debates,* eds. B Capps and A Campbell. In press: World Scientific/Imperial College London, 2010.

[31] Warnock M, *Report of the Committee of Enquiry Into Human Fertilisation and Embryology,* UK Committee on Human Fertilisation and Embryology, 1984.

[32] National Bioethics Advisory Commission, USA, *Ethical Issues in Human Stem Cell Research,* 1999.

scientific jurisdictions. The policies of the UK, US, Canada, and Australia were especially important due to historical factors and institutional affiliations, as well as broadly similar political cultures. As the Biomedical Sciences Initiative was understood from the start to be a long-term investment, it was critical for Singapore not to stand out as an apparent 'rogue state'. However, as Richard Magnus observed, a consequence of this approach is that the BAC's recommendations may come across as having a 'Western' character, even though there are distinctively 'local' features, such as the recommendation by the BAC for no one to be "under a duty to participate in any manner of research on human stem cells, which would be authorised or permitted by the law, to which he has a conscientious objection".[33]

Another important development that was initiated from the first report of the BAC was the progressive promulgation of a set of ethical principles or 'goals' that should be the basis of ethical review of biomedical research in Singapore. This case-by-case development of the ethical framework resembles the approach of the UK Nuffield Council on Bioethics.[34] Five ethical principles (justice, sustainability, respect for individuals, reciprocity, and proportionality) have been restated in their entirety in the BAC's 2008 report on egg donation (2008 restatement).[35] They may arguably be viewed as representative of ethical thinking in Singapore, in much the same way as the Belmont principles (respect for persons, beneficence, and justice)[36] or the four ethical principles (respect for autonomy, beneficence, non-maleficence, and justice) of Beauchamp and Childress[37] have come to characterise American bioethics.

Two of the five ethical principles, 'just' and 'sustainable', were set out in the first report. However, the ethical content of 'just' and 'sustainable' set out in relation to stem cell research and cloning is broader than that

[33] Bioethics Advisory Committee, Singapore, *Ethical, Legal and Social Issues in Human Stem Cell Research, Reproductive and Therapeutic cloning*, June 2002, p. 35, Recommendation 11.
[34] The BAC is similar to the Nuffield Council on Bioethics in a number of ways. Like the Council, the BAC is also not bound by the values of particular schools of philosophy or approaches in bioethics. See Whittall H. A Closer Look at the Nuffield Council on Bioethics, *Clinical Ethics* **3** (2008): 199–204.
[35] Bioethics Advisory Committee, Singapore, *Donation of Human Eggs for Research*, November 2008, p. 10, para 4.1.
[36] The National Commission for the Protection of Human Subjects of Biomedical and Behavioral Research, USA, *The Belmont Report: Ethical Principles and Guidelines for the Protection of Human Subjects of Research*, 18 April 1979.
[37] Beauchamp TL and Childress JF. *Principles of Biomedical Ethics*. New York: Oxford University Press, 2008 (6th ed).

of the restatement as they also relate to a policy orientation of the BAC. The principles of 'just' and 'sustainable' in the former context relates to fair representation and viable course of action. Sociologist Eddie Kuo observed that in Singapore, consensus may be difficult to arrive at given the diversity in religion, ethnicity, and culture that constitutes the Singaporean public. This challenge was similarly acknowledged by many of the past and present BAC members who were interviewed. Despite this, Tan Chorh Chuan indicated that as a diverse society, open consultation can be a constructive and inclusive platform for fair representation of different viewpoints in the interest of fostering better understanding of these perspectives and of building consensus.

The right to conscientious objection was adopted out of respect for persons and leaves some leeway for certain groups that are opposed to the research. BAC Chairman Lim Pin similarly observed that Singapore is a diverse society and hence an empathic, practical, and realistic approach is often necessary to avoid a standstill and social stagnation. On the stem cell debate, he recognised that those opposed to the research are sincere and not motivated by self-interest. However, as there is no single dominant view on the subject in Singapore, it was necessary for the BAC to craft a *sustainable* position even if not everyone will agree, provided that they have had a *fair* opportunity to express their views. In other words, the BAC has been a critical intermediary in bringing together views, in some cases irreconcilable, of different groups and communities. Hence, the principles of 'just' and 'sustainable' as presented in the context of stem cell research and cloning related to a public policy orientation of the BAC, whereas the 2008 restatement of these principles are directed more specifically at research, so that 'just' is concerned with the fair distribution of the benefits and burdens of research, whereas 'sustainable' is directed at the impact of research on future generations.[38]

As it turned out, a 14-day limit on the development of embryos for research, coupled with regulatory control and the right to conscientious objection, constituted a plausible passage through an ethical impasse. The

[38] Bioethics Advisory Committee, Singapore, *Donation of Human Eggs for Research,* November 2008, p. 10, para 4.1 (d) and (e). The ethical principles of 'just' and 'sustainable' thereby follow from the BAC's narrower reading of 'just' as allowing "research with tremendous potential therapeutic benefits to mankind" to proceed and 'sustainable' as producing research with "little biological or genetic impact on future generations". See Bioethics Advisory Committee, Singapore, *Ethical, Legal and Social Issues in Human Stem Cell Research, Reproductive and Therapeutic Cloning,* June 2002, p. 35, para 47.

recommendations of the BAC include proposals for stringent regulation of human embryonic stem cell research in Singapore and the legal prohibition of reproductive cloning, which was taken up by the legislature with the enactment of the *Human Cloning and other Prohibited Practices Act* in 2004.[39] For research that involves the creation of human embryos (such as therapeutic cloning), specific regulatory approval is required. In addition, the research must also be justified by strong scientific merit and potential medical benefit.[40]

Following the publication of these recommendations, scientific developments in relation to cloning and induced pluripotent stem cell technologies necessitated a review of the recommendations. This was formally undertaken by the HECR Working Group in 2007, with a focus on the ethical, legal, and social issues arising from the procurement and use of human eggs for biomedical research, and on research involving human-animal combinations. Apart from scientific developments, review of these areas was considered necessary following the scandal involving unethical procurement of human eggs for research in South Korea[41] and, more importantly, revisions to ethical policies and guidelines in the United States, Australia, Canada, and a number of European countries such as Britain and Denmark. This initiative was also undertaken by the BAC as part of its longer-term intention to consolidate its views and recommendations in the area of human embryonic stem cell research.[42] It was apparent that there was increasing pressure to find a sustainable source of eggs for research, particularly for stem cell science, where the scarcity of human eggs is the limiting factor for therapeutic cloning. One solution is to use animal eggs, which became part of the public consultation on human-animal combinations conducted immediately following the closure of the consultation on egg donation.[43] Another alternative is to increase the number of eggs donated by women. One of the issues this raised was the possibility of offering incentives to donate for research purposes. This introduced

[39] Singapore Statutes: *Human Cloning and Other Prohibited Practices Act* (Cap. 131B), Revised 2005.
[40] Bioethics Advisory Committee, Singapore, *Ethical, Legal and Social Issues in Human Stem Cell Research, Reproductive and Therapeutic Cloning,* June 2002, pp. 27–29.
[41] Cyranoski D. Korea's Stem-Cell Stars Dogged by Suspicion of Ethical Breach, *Nature* **429**, 6987 (2004): 3.
[42] Bioethics Advisory Committee, Singapore, Press Release, 7 November 2007, para 2.
[43] Bioethics Advisory Committee, Singapore, Press Release, 8 January 2008. The late Anne McLaren indicated that human eggs are scarce and not usually available for research: McLaren A. Free-Range Eggs? *Science,* **316**, 5823 (2007): 339.

not only the issue of egg trading, but also the possibility of inducing women into a potentially risky and invasive procedure of ovarian stimulation and egg collection with no direct therapeutic benefit.[44]

Recommendations relating to the donation of human eggs for biomedical research were published by the BAC at the end of 2008, after public feedback was received on various issues presented in a consultation paper[45] and at a public forum on 11 November 2007. In addressing the issue of egg donation, the BAC took the opportunity to clarify the scope of the 'non-commercialisation of the human body' requirement under the principle of respect for individuals. This principle was earlier applied in the context of donation of human tissue for research, where the BAC indicated that the donor should relinquish any property or property-like claims over the 'gifted' tissue:[46]

> *Although a donor may make an outright gift of his or her tissue in the sense [that] he or she renounces any property rights to or in connection with the tissue, it is entirely open to the donor to stipulate or define the kind of research uses to which the tissue may be applied.*

In its report on egg donation, the BAC recommended that women donating eggs for research should be reimbursed for expenses incurred and compensated for loss of time and earnings as a result of the procedures required to obtain the eggs.[47] Non-commercialisation of eggs was emphasised, as this was regarded as necessary to avoid putting women at risk of exploitation.[48] This is consistent with the BAC's goal of safeguarding the welfare of all research participants. Should an egg donor suffer from any medical complication as a direct and proximate result of the donation, she should be provided with prompt and full medical care. This provision gives effect to public feedback on the need to ensure that medical care is available for adverse

---

[44] Bioethics Advisory Committee, Singapore, *Donation of Human Eggs for Research*, November 2008, p. 4, para 1.3.

[45] Bioethics Advisory Committee, Singapore, *Donation of Human Eggs for Research: A Consultation Paper*, 7 November 2007.

[46] Bioethics Advisory Committee, Singapore, *Human Tissue Research*, November 2002, p. 24, para 8.6.

[47] Bioethics Advisory Committee, Singapore, *Donation of Human Eggs for Research*, November 2008, p. 3, Recommendation 6.

[48] *Ibid.* pp. 16 and 17, paras 4.16–4.21.

health consequences arising from the egg donation procedure. Responsibility for this provision rests with the researchers and their institutions.[49]

The BAC maintains that the ethical requirement for the donation of tissue (which includes embryos) for research to be outright gifts is not compromised so long as the contribution is not tainted by any inducement.[50] The giving of eggs for research is still altruistic if compensation that is directed at ensuring the financial neutrality of the contributor does not amount to an inducement. In other words, it is consistent with the principle of justice to allow compensation to be provided for loss of time and earnings that are consequential to the donation. In contrast, women should not be compensated for the donation of eggs for research when these are surplus to the treatment or obtained as a result of other medical treatments. No additional discomfort or inconvenience would have been assumed by these women as the risk, discomfort, and lost time are already an inherent part of the treatment.[51] Hence, the BAC did not consider the 'compensated egg sharing' schemes adopted in the UK to be acceptable in Singapore.[52]

The BAC took the view that respect for the human body is fundamental to ethical thinking and conduct in both medical practice and biomedical research.[53] Commercialisation of the human body, by treating it, or part of it, as a disposable economic asset is generally taken to be inconsistent with this principle. It noted that this view is not unchallenged, but insofar as it underpins current ethical thinking in Singapore, it supports a view that financial inducement to provide tissues or cells for research would amount to a form of commercialisation and is not acceptable.[54] This view found support among public institutions and the general public.[55]

---

[49] *Ibid.* p. 3, Recommendation 5.
[50] *Ibid.* p. 16, para 4.18, reiterating the position taken in *Human Tissue Research*, November 2002, pp. 35–36, paras 13.1.8–13.1.10.
[51] *Ibid.* p. 22, para 4.28.
[52] *Ibid.* pp. 20–21, paras 4.23–4.24. On the position in the UK, see Capps B and Campbell A. Why (only some) Compensation for Oocyte Donation for Research Makes Ethical Sense, *Journal of International Biotechnology Law* **4** (2007): 89–102.
[53] Bioethics Advisory Committee, Singapore, *Donation of Human Eggs for Research*, November 2008, p. 16, para 4.16.
[54] Campbell AV. *The Body in Bioethics*. Oxford and New York: Routledge-Cavendish, 2009. See also Ho WC, Capps B and Voo TC. Stem Cell Science and its Public: The Case of Singapore, *East Asian Science, Technology and Society: an International Journal* **39**, 1 (2010): (in press).
[55] See feedback from public consultation published with the report: Bioethics Advisory Committee, Singapore, *Donation of Human Eggs for Research*, November 2008, pp. C-2 to C-111. No respondent indicated support for the sale of human eggs.

In January 2008, a consultation paper on research involving human-animal combinations was distributed for public discussion and comment.[56] As with the consultation paper on egg donation, but unlike other consultation papers previously issued by the BAC, this paper did not propose any recommendations and thus had a more open-ended character. BAC member Nazirudin Mohd Nasir welcomed this approach as it presented the BAC as initiating consultation without committing to any ethical position.[57] HECR Working Group Chairman Richard Magnus[58] and member Eddie Kuo[59] were concerned that the subject of human-animal combinations is too broad and complex, thereby compounding the difficulty in assessing its ethical and social implications. Stem cell researcher and member of the HECR Working Group, Lim Bing,[60] further indicated that research involving human-animal combinations could be technically challenging even for IRB members. Hence, there may be a need for a specialist ethics body to be established for the review of such research at a national level. Although human-animal combinations is complicated and may not be a matter of public concern, Ng Soon Chye stated that it would not be sensible to delay consideration as human-animal combinations are increasingly commonplace constructs in biomedical research, and it is important that scientists know the ethical boundaries.[61] Recommendations on this subject are still being debated and deliberated on by the HECR Working Group and the BAC.

RECOMMENDATIONS IN RELATION TO RESEARCH INVOLVING HUMAN SUBJECTS AND GENETICS

The HGS was responsible for a series of recommendations which served to systematise ethical governance of research using human tissue, research involving human subjects, and genetics research. These recommendations were published in four reports. HGS Chairman Terry Kaan[62] said that the topics considered were drawn from a broad review of ethical, policy, and regulatory debates around the world, and in consultation with the local medical and research communities. The need for a set of national guidelines on

[56] Bioethics Advisory Committee, Singapore, *Human-Animal Combinations for Biomedical Research: A Consultation Paper*, 8 January 2008.
[57] Interview with Mr Nazirudin Mohd Nasir, 1 April 2010.
[58] Interview with Mr Richard Magnus, 18 April 2009.
[59] Interview with Professor Eddie Kuo, 28 April 2009.
[60] Interview with Dr Lim Bing, 13 July 2009.
[61] Interview with Professor Ng Soon Chye, 26 May 2009.
[62] Interview with Associate Professor Terry Kaan, 16 June 2009.

the ethical derivation and use of human tissue was a pressing concern at that time as human tissue is a fundamental resource for most biomedical research. Fortunately, the HGS did not have to 're-invent the wheel' as many medical practitioners and researchers, especially those with some level of training in leading scientific jurisdictions, would already be familiar with ethical requirements. In many instances, the HGS focused on developing a set of ethical practices that was best suited to local conditions by drawing from the best ethical practices in key jurisdictions. The task was no less challenging as it was not always easy to decide which of the available ethical best practices is most suitable, particularly if the immediate objectives of medical practitioners and researchers do not converge.

The report on human tissue research, published in November 2002, provides a set of national ethical guidelines to be applied uniformly to all persons involved in human tissue banking and research using human tissue in Singapore. The ethical principles (referred to as governing ethical principles) embodied in the guidelines have been set out as the primacy of the welfare of tissue donors, the need for informed consent, confidentiality and ethics review, respect for the human body, and sensitivity towards the religious and cultural perspectives and traditions of tissue donors.[63] As earlier considered, the ethical requirement of non-commercialisation follows from the principles of respect for the human body and the primacy of the welfare of tissue donors. All the governing ethical principles are in turn encapsulated within the principle of respect for individuals under the 2008 restatement of the BAC's general ethical principles. Kaan said that once a set of ethical guidelines for human tissue research was developed, it then became necessary to consider how bioethical oversight was to be carried out. An ideal ethics governance structure should be enabling, so that researchers and their IRBs can work through differences.

In November 2004, a set of guidelines for IRBs on research involving human subjects was published to formalise the requirement for all human biomedical research in Singapore, including research involving human tissue or medical information, to be subject to ethics review. These guidelines built on the existing system of regulations for pharmaceutical trials and human biomedical research conducted by hospitals, private clinics, and other healthcare establishments under the supervision of the MOH. The NMEC's

[63] Bioethics Advisory Committee, Singapore, *Human Tissue Research*, November 2002, pp. 33–36, para 13.1.

guidelines on research involving human subjects[64] and the Singapore Guideline for Good Clinical Practice[65] were the most influential. The BAC's IRB guidelines[66] set out the constitution, accreditation, and operation of IRBs, as well as the roles and responsibilities of IRBs, research institutions, and individual researchers. The normative justification for the structure of ethical governance is in turn grounded in internationally accepted values promulgated in documents such as the *Declaration of Helsinki*, the *Nuremberg Code*, the *Belmont Report*, and UNESCO's *Universal Declaration on the Human Genome and Human Rights*.[67] In the main, the ethical principles[68] that underlie the structure of ethical governance entailed in the IRB Guidelines (and in addition to those already specified by the NMEC) are captured by the principle of respect for individuals under the 2008 restatement. Also implicit in the structure is the principle of proportionality, most evidently reflected in the different levels of ethics review (exempted, expedited, or full review). All human biomedical research, which "involves any direct interference or interaction with the physical body of a human subject, and that involves a concomitant risk of physical injury or harm, however remote or minor"[69] should be "... [fully] reviewed and approved by a properly constituted ethics committee or IRB".[70] However, research using established commercially available cell lines or commercially available and anonymised human material or research involving the analysis of patients' information but without any interaction with patients may be exempted from ethics review or qualify for expedited review.[71]

The principles of reciprocity and proportionality are explicitly set out and explained in the BAC's deliberations on genetic research. Ethical governance

---

[64] National Medical Ethics Committee, Singapore, *Ethical Guidelines on Research Involving Human Subjects*, August 1997.

[65] Ministry of Health, Singapore, *Singapore Guideline for Good Clinical Practice*, 1998, Revised 1999. This document has the regulatory effect under Section 21 of the *Medicines (Clinical Trials) (Amendment) Regulations*, 2000 Revised Edition.

[66] Bioethics Advisory Committee, Singapore, *Research Involving Human Subjects: Guidelines for IRBs*, November 2004.

[67] *Ibid.* p. 25, para 4.8.

[68] *Ibid.* p. 27, para 4.17. These ethical principles are respect for the human body, welfare and safety, and for religious and cultural perspectives and traditions of human subjects, respect for free and informed consent, respect for privacy and confidentiality, respect for vulnerable persons, and avoidance of conflicts of interest or the appearance of conflicts of interest.

[69] *Ibid.* p. 17, para 3.7 (a).

[70] *Ibid.* p. 18, para 3.9.

[71] *Ibid.* p. 19, para 3.15 (c), and p. 20, para 3.18 (c).

of genetic research was formulated to apply at two different junctures: at the point where genetic information is derived through various means of testing, and in the management and use of the information itself. The report on genetic testing and genetic research served to operationalise a number of internationally recognised ethical considerations in the local context. These ethical considerations relate essentially to the principle of respect for individuals, as they require respect for the welfare, safety, and religious and cultural perspectives and traditions of individuals; informed consent; respect for vulnerable persons; and privacy and confidentiality of genetic information.[72] Specific ethical considerations have also been set out by the BAC in relation to five types of genetic testing and it further recommended that non-consensual or deceitful taking of human tissue for the purpose of genetic testing be prohibited by law.[73]

There are at least two features that point to a less individualistic (and perhaps more communitarian) approach to decision-making for genetic testing. First, genetic testing is by and large regarded as a medical procedure. Consequently, genetic information is seen effectively as medical information even though its accessibility to certain third parties (such as insurers and employers) may be more limited than general medical information.[74] Second, the BAC agrees with the World Health Organization that pre-symptomatic or susceptibility testing of children and adolescents should be carried out only if there are potential medical benefits to them.[75] However, in a specific application of carrier testing, the BAC recognises that it would be ethical for such testing to be done if the family concerned could benefit from this knowledge:

> *We recognise that as a matter of principle, carrier testing in asymptomatic children should generally be deferred until the child is mature or required to make reproductive decisions. This is because to do otherwise is to risk pre-empting a later decision by the child, when adult, not to know his or her own genetic status or have it made known to others ... However, the defence of this right*

[72] Bioethics Advisory Committee, Singapore, *Genetic Testing and Genetic Research*, November 2005, p. 23, para 4.1.
[73] *Ibid.* pp. 30–41, and p. 9, Recommendation 4.
[74] *Ibid.* p. 22, Recommendation 1.
[75] World Health Organization, *Proposed International Guidelines on Ethical Issues in Medical Genetics and Genetic Services*, 1998, p. 9, Table 6.

> *must be weighed against the interests of other family members, the proper medical care of whom may depend on full and accurate information about a genetic condition in the family, as well as the wider public health interests of a given community. In Singapore, genetic screening programmes for at-risk groups aimed at lowering the incidence of lethal or disabling genetic conditions common in the local population, such as thalassaemia, are widely supported by both the medical profession and the public... Where compelling interests of other family members or public health exist, we are of the view that the physician should be able to decide, together with the parents, whether or not to determine the carrier status of the child... (emphasis added)*[76]

The social and ideological significance of communal interest was the basis for the principle of reciprocity in the BAC's report on personal information. This principle was explained as:

> *...the idea that accepting benefit from past medical research, inherent in the utilisation of medical services, carries some expectation of a willingness to participate in research for the common good or public interest. This is an especially important consideration in societies, including Singapore society, where individuals are seen as incurring obligations to others through their membership of and roles in society. In the wider public interest, therefore, we see the principles of autonomy and reciprocity as complementary.*[77]

Nuyen Anh Tuan explained[78] that reciprocity presupposes individuality in that each agent is seen as independent and autonomous. These agents will co-operate with one another if it serves their interests, on a "you scratch my back, I scratch yours" rationale. Hence, reciprocity is entailed in co-operation. In contrast, the principle of solidarity presupposes a community

[76] Bioethics Advisory Committee, Singapore, *Genetic Testing and Genetic Research*, November 2005, p. 26, para 4.13.
[77] Bioethics Advisory Committee, Singapore, *Personal Information in Biomedical Research*, May 2007, p. 27, para 5.19.
[78] Interview with Associate Professor Nuyen Anh Tuan, 21 April 2009.

wherein agents have particular social roles based on their positions in the community. In this respect, solidarity is similar to Confucianism,[79] and co-operation is undertaken for the preservation of the community. Although reciprocity and solidarity are conceptually different, their net effect is the same in that the interests of a community are served.[80] Nuyen considered the BAC to have brought the notions of individualism (in reciprocity) closer to solidarity with its indication that reciprocity is not just an exercise of self-interest, but a duty to contribute a return to the community under a sense of continuity as involving future generations and having benefited from past generations.[81] Ethical justification grounded in the principle of reciprocity was applied to the recommendation of the BAC to provide firm legal footing to disease registries that employ personal information in public health research.[82] This recommendation contributed to the enactment of the *National Registry of Diseases Act* later that year.[83]

Also explicated in its report on the use and management of personal information in biomedical research is the principle of proportionality *vis-à-vis* informed consent (or more generally, autonomy). It indicated that: "the process of obtaining consent should be detailed in proportion to the sensitivity of the research and the actual or perceived risk of harm to the individual concerned".[84] It followed that general consent would suffice for research using de-identified information, but specific consent is required for the types of genetic research that may be of public concern, such as those relating to personality, behavioural characteristics, sexual orientation, and intelligence.[85] The application of this principle is further evident in the BAC's discussion on

[79]Nuyen AT. Moral Obligations and Moral Motivation in Confucian Role-Based Ethics, *Dao* **8** (2009): 1–11.
[80]For a discussion on the ethical principles of reciprocity, solidarity, mutuality, citizenry, and universality, see Knoppers BM and Chadwick R. Human Genetics Research: Emerging Trends in Ethics, *Nature Reviews: Genetics* **6** (2005): 75–79.
[81]The BAC explained that "Existing patients are receiving the benefits of improved medical care through the use of medical information from past patients for research. There is little ethical justification for them to refuse a similar use of their medical information where their interests are not likely to be compromised." See Bioethics Advisory Committee, Singapore, *Personal Information in Biomedical Research*, May 2007, p. 30, para 5.25.
[82]*Ibid.* p. 33, Recommendation 3.
[83]Singapore Statutes: *National Registry of Diseases Act* (Cap. 201B), Revised 2008.
[84]Bioethics Advisory Committee, Singapore, *Personal Information in Biomedical Research*, May 2007, p. 25, para 5.13.
[85]*Ibid.* p. 25, para 5.11.

de-identification and the accessibility of personal information by third parties such as employers and insurers.[86]

THE VALUE OF BIOETHICS

The public has had a part to play in shaping the ethical framework put forward by the BAC. The BAC's recommendations are the outcome of professional and, in almost every case, extensive public feedback and suggestions. All public consultations have been widely publicised by the local media in order to encourage public deliberation and participation, and at least one public forum would be organised. In a number of cases, special meetings with religious group leaders and researchers were convened. The BAC has also worked closely with REACH (Reaching Everyone for Active Citizenry @ Home), the lead agency for engaging and connecting with citizens of Singapore. Previously known as the Feedback Unit, it has been tasked by the government to encourage and promote an active citizenry through citizen participation and involvement. Although the work of the BAC does not directly relate to government policies (such as tax, transportation, or education), Toh Yong Chuan and Lilian Ong of REACH said that it shares the BAC's interest in enabling the public to engage with bioethical issues and in gathering feedback on these issues.[87] In this connection, REACH has assisted the BAC in facilitating public discussion on bioethical issues and gathering public feedback. Although REACH has over 10,000 people on their contact list, including Singaporeans who are overseas, Toh and Ong indicated that bioethics is not a subject that many are naturally interested in. They added that when bioethical issues are presented to the public, it is important to be clear about the 'target audience', such as making a distinction between issues that affect the general public and those that are of interest only to researchers.

How much public interest in bioethics can we expect? Cheong Yip Seng, former Chair of the PES, said that it would be difficult to interest every member of society in bioethics given its technical nature. In addition, the 'man in the street' might not see the need to get involved as the Singaporean public has a high regard for science and especially learning.[88] There would not

[86] *Ibid.* pp. 20–22 (on de-identification) and pp. 38–41 (on access to medical information by employers and insurers).
[87] Interview with Mr Toh Yong Chuan and Ms Lilian Ong, 19 June 2009.
[88] Interview with Mr Cheong Yip Seng, 9 April 2009.

ordinarily be a need to question the way in which science and technology are managed in Singapore, much in the way that the general public would not be too concerned with how the Monetary Authority of Singapore manages the financial system. However, he felt it would still be important for the BAC to frame issues in a manner that could involve as much of the public as possible. He considered the BAC's website and the media to be important means by which the BAC could engage the public. This engagement is important because it gives legitimacy to the research and makes clear what the ethical expectations are for both the public and researchers. He added that when he was editor of *The Straits Times* (the main English language newspaper in Singapore), a science section was introduced to keep its readers informed of recent scientific advancements.

Han Fook Kwang, who took over from Cheong Yip Seng as Chair of the PES, shared a similar view.[89] He said that most Singaporeans have little knowledge of the BAC and its work, as it does not deal with 'bread-and-butter' issues. He added that Singaporeans are generally very practical and have to cope with "a very noisy world out there. Bioethics could get 30 seconds of fame with interesting personalities or issues, but it is not a daily news item." He also agreed with Mr Cheong that the public has a fairly positive view of the government and its ability, and hence may not see the need to get involved. However, public engagement is still important to ensure that the public understands (or does not misunderstand) the significance of investing in biomedical sciences.[90] He added that there has fortunately been no major issues so far but would not rule out a potential 'blow-up' if there was misunderstanding over a particular issue. He was not sure if it was meaningful for the BAC to try to engage the public in a general way over bioethical issues. The best approach was to do so when there are specific issues that need

[89] Interview with Mr Han Fook Kwang, 31 August 2009.

[90] See Skidelsky R. The Price of Clarity, *The Straits Times*, 24 May 2010, p. A24. Robert Skidelsky's remark, although made in relation to the financial crisis triggered by the US subprime mortgage situation, is apt here:

> *The greater the distance between the language of elites and ordinary people, the greater the risk of revolt. To the extent that complexity in finance or politics creates new opportunities to deceive or impedes understanding, we should aim to reduce it. To the extent that such problems reflect decreased ability to express oneself clearly, the remedy is to improve education. The price of clarity, like the price of liberty, is eternal vigilance, and the two are connected.*

public understanding and support, and to canvass a wide range of views in these instances, including those from the non-English speaking public.

As Harry Collins and Robert Evans observed, the concept of expertise has become so deeply woven into shared understanding that it would be difficult to arrive at particular goals without the benefit of expert input.[91] The BAC has become a means by which bioethical issues are considered and addressed in Singapore. Charles Lim considered the deliberation of the BAC to comprise robust discussions by which different views were canvassed, debated, and distilled until a common ground was reached. Even then, this position might not necessarily be one of full consensus.[92] Through the years, healthcare and research institutions, governmental entities, and professional and religious organisations have been in consultation with the BAC on its recommendations. Depending on the issue at hand, individual lay members of society (especially students) have also been engaged. The relationship between the BAC and the public has redefined itself in ways that could be limiting and enabling.[93] The notions of 'public' and 'expertise' are difficult to define, leading some to conclude that bioethics is not a distinct discipline, but a consortium of different expertise.[94]

Even then, bioethics as a kind of 'placeholder' has social value.[95] Peter Singer argues that scientists (and bioethicists) do not have any greater access to 'truth', or otherwise any claim to special information or reasoning ability capable of devising a perfect and undisputed moral code.[96] However, he argues that thinking through an issue — especially by socially responsible people — and arriving at a "soundly based conclusion"[97] is difficult and time-consuming because it requires gathering detailed and reliable information, thorough analyses and assessment of the information within one's moral

[91] Collins HM and Evans R. The Third Wave of Science Studies: Studies of Expertise and Experience, in *The Philosophy of Expertise*, eds. E Selinger and RP Crease. New York: Columbia University Press, 2006, pp. 39–110. See also Jasanoff S. *Designs on Nature: Science and Democracy in Europe and the United States*. Princeton: Princeton University Press, 2005, p. 250.
[92] Interview with Mr Charles Lim, 14 January 2008.
[93] Riles A. *The Network Inside Out*. Michigan: University of Michigan Press, 2001, pp. 58–59.
[94] O'Neill O. *Autonomy and Trust in Bioethics*. Cambridge: Cambridge University Press, 2002, p. 1.
[95] Riles A. *Collateral Knowledge: Legal Reasoning in the Global Financial Markets*. In press, Chapter 5.
[96] Singer P. Moral Experts, in *The Philosophy of Expertise*, eds. E Selinger and RP Crease. New York: Columbia University Press, 2006, pp. 187–189.
[97] *Ibid*. p. 189.

views, while guarding against bias. Bioethics, as a form of moral (albeit collective) expertise, is better able to accomplish this task. Yap Hui Kim expressed a similar view.[98] She indicated that public bodies like the BAC need responsible people with views that are not skewed or self-interested. She accepted that a majority of the public would be apathetic but felt that the BAC must engage with individuals, groups, and institutions (like think tanks) who are able to invest the time and resources to think through and speak in the interest of the common good. It is also such engagement that provides a reason for some individuals and organisations to invest time and resources in understanding bioethical issues. Nazirudin Mohd Nasir indicated that his organisation (the Islamic Religious Council of Singapore) further built up its resources on bioethics as an outcome of its engagement with the BAC.[99] He agreed with Member of Parliament Zainul Abidin Rasheed that social institutions could adapt to technological changes, such as the reliance on astronomical calculations when the actual sighting of the moon is not possible, in the determination of when the fast of Ramadan (an important religious festival) begins.[100]

Bioethics presents (and represents) the collective fabric of society and is a means of achieving what Zainul Abidin described as "secularism with a soul".[101] To varying degrees, bioethics reconciles science with religion on a normative platform, so that the agents of both enterprises are answerable to the collective. This observation of Fox and Swazey appears to support these views:

> *..."using biology and medicine as a metaphorical language and a symbolic medium, bioethics deals in public spheres and in more private domains with nothing less than beliefs, values, and norms that are basic to our society, its cultural tradition, and its collective conscience"... While recognizing the*

[98] Interview with Professor Yap Hui Kim, 20 April 2009. She indicated that it is important that there is fair representation and no single viewpoint should dominate. Public participation is an effective way to avoid hegemony.

[99] Interview with Mr Nazirudin Mohd Nasir, 1 April 2010.

[100] The use of astronomical calculations in the determination of when religious festivals begin is by no means a settled issue for Muslim communities in other parts of the world. See Shah ZA. *The Astronomical Calculations and Ramadan: A Fiqhi Discourse*. Washington and London: International Institute of Islamic Thought, 2009.

[101] Interview with Member of Parliament Mr Zainul Abidin Rasheed, 23 July 2009.

> *basic interconnection of bioethics to advances in modern biology, medicine, and biotechnology, we have always been impressed by the degree to which "the value and belief questions with which [the field] has been preoccupied have run parallel to those with which the society has been grappling more broadly," and by their wider "moral, social, and religious connotations"...* [102]

To some extent, bioethics may appear to serve as a form of 'public relations' for researchers. The importance of public trust is generally recognised by the research community, as Lim Bing indicated,[103] and the public cannot be forced to accept scientific advances that it considers objectionable. However, Zainul Abidin indicated that the work of the BAC was not intended to be a public relations exercise. There has been genuine interest on the part of the BAC to engage meaningfully with the public in advancing the Biomedical Sciences Initiative. This initiative will be met with social resistance unless the public is genuinely convinced that the science is of value to society.[104] Alastair Campbell went further, indicating that bioethics has a more fundamental role in setting out ethical principles or values that all members of society are answerable to, whether researcher or otherwise. Hence, the BAC is not merely a public relations agency or a feedback unit. It steers (collaboratively with other bioethical bodies) the long-term course of the scientific enterprise through the definition of standards and norms (both local and international) as to the measure of 'good science'.

A process by which the BAC derived its ethical principles was by achieving consensus through intermediation.[105] Richard Magnus (Chair of the HSCRS and the HECR Working Group)[106] and Terry Kaan (Chair of the HGS)[107] have both indicated that intermediation is akin to adjudication, or the manner

[102] Fox RC and Swazey JP (with the assistance of Watkins JC). *Observing Bioethics*. New York: Oxford University Press, 2008, pp. 6–7.
[103] Interview with Dr Lim Bing, 13 July 2009.
[104] This is perhaps all the more so since the risk of the investment is borne by the population: Waldby C. Singapore Biopolis: Bare Life in the City-State, *East Asian Science, Technology and Society: An International Journal* **3** (2009): 367–383, pp. 381–382.
[105] Pielke RA Jr. *The Honest Broker: Making Sense of Science in Policy and Politics*. Cambridge: Cambridge University Press, 2007. Pielke sets out different types of intermediation or 'brokerage' in a science policy environment.
[106] Interview with Mr Richard Magnus, 10 April 2009.
[107] Interview with Associate Professor Terry Kaan, 16 June 2009.

by which a court of law resolves competing claims.[108] Other legal experts on the BAC, Jeffrey Chan and Charles Lim,[109] indicate the range of possible outcomes from ethical deliberation to be significantly broader than adjudication. More importantly, all four legal experts have indicated that values such as independence, transparency, and balance that are critical to a sound judicial system are similarly vital to the BAC.[110] An observation could be made here that although bioethics may lack a dominant methodology or discourse, it is not devoid of (at least in the case of a public bioethical body like the BAC) normative content. The extent of intermediation was not limited to differences between the public and the research community, but also included those between the research community and different expert or professional communities (such as scientific, medical, and regulatory), as well as within the research community. Jeffrey Chan observed that there were often differences of views between practitioners or researchers (who focused on outcomes and thus desired flexibility) and regulators (who sought to impose restrictions based on public policy). Mechanisms must be provided to balance these competing considerations. His preference was for scientific governance through clear legislative provisions rather than through ethical preferences of the moment. Clear laws make for a more certain and transparent system.[111]

However, Edison Liu was concerned that too many legal provisions could burden research, especially if IRBs should operate in a legalistic manner (or if "ethics becomes law").[112] Chia Kee Seng was similarly concerned.[113] He indicated that if regulators or IRBs fear liability and adopt a "letter of the law" mentality (ie. applying ethical rules strictly and without flexibility in accommodating exceptional circumstances), research will be impeded. If society values research, there should be an authority or body that could speak up

---

[108]Writing in relation to health care ethics committees in the US, Diane Hoffmann and Anita Tarzian indicate that these committees serve as a mechanism for dispute resolution for issues that arise out of medical practice in many states. See Hoffmann DE and Tarzian AJ. The Role and Legal Status of Health Care Ethics Committees in the United States, in *Legal Perspectives in Bioethics*, eds. AS Iltis, SH Johnson and BA Hinze. New York and London: Routledge, 2008, pp. 46–67, pp. 46–47.
[109]Interview with Mr Jeffrey Chan Wah Teck SC, 13 April 2009; and Interview with Mr Charles Lim, 14 January 2008.
[110]Interestingly, these values have similarly been identified to be critical to national bioethics advisory bodies in the US: Briggle A. The Kass Council and the Politicization of Ethics Advice, *Social Studies of Science* **39**, 2 (2009): 309–326.
[111]Interview with Mr Jeffrey Chan Wah Teck SC, 13 April 2009.
[112]Interview with Professor Edison Liu, 8 July 2009.
[113]Interview with Professor Chia Kee Seng, 6 April 2009.

for researchers. It is inevitable — like collateral damage in war — that some researchers may stray into unethical territory from time to time. But any disciplinary action that follows must be edifying and not destructive. He allegorised this as mentoring a child. He emphasised the need to help researchers resolve ethical dilemmas in a constructive manner, a point that Yap Hui Kim also made.[114] She said that there would be an increasing number of gray areas with rapid advances in biomedical sciences. A researcher might also be confused by different ethical standards that IRBs apply. She was concerned that one IRB may be more lax than another, hence some effort at harmonisation is necessary.[115] Lim Bing expressed the same concern,[116] indicating that the quality of ethics review differs among various IRBs in Singapore. An IRB composed of vocal members who are not well informed about science could hinder research, in addition to the burden of bureaucratic paperwork entailed in the ethics review process. Hence, the BAC has a role in harmonising the various current ethical practices based on the guidelines and practices of different leading jurisdictions — all acceptable to varying degrees and circumstances — that have been adopted by a diverse group of highly trained researchers, medical professionals, and IRB administrators. This role, as Terry Kaan indicated, was already taken up by the BAC when it deliberated on ethical practices for the use of human tissue in research, and will continue to be a critical function of the BAC.[117]

Aside from its 'active' role in engaging with and intermediating among many different levels and groups or segments of society, the BAC also has a more passive role that its Chairman Lim Pin described as akin to empathic listening.[118] He said that listening is important as a form of empathy. From his experience, it took away a lot of anger and frustration. Ng Soon Chye, who has been involved in a number of public sessions organised by the BAC, said that some people attend these sessions or lectures not to learn, but to express

[114] Interview with Professor Yap Hui Kim, 20 April 2009.

[115] In the US, lack of homogeneity in structure and operation among health care ethics committees was attributed in part to the absence of federal regulation. Hence, the role and legal status of ethics committees in the US were regarded as "amorphous". See Hoffmann DE and Tarzian AJ. The Role and Legal Status of Health Care Ethics Committees in the United States, in *Legal Perspectives in Bioethics*, eds. AS Iltis, SH Johnson and BA Hinze. New York and London: Routledge, 2008, pp. 46–67, pp. 63.

[116] Interview with Dr Lim Bing, 13 July 2009.

[117] Interview with Associate Professor Terry Kaan, 16 June 2009.

[118] Interview with Professor Lim Pin, 22 April 2009.

their views.[119] An important point to be noted here is that although public sessions or lectures are often viewed as a means of 'public education', or for the skeptics, 'public indoctrination', past experience suggests that these occasions have also been important for public expression. Public feedback, including views that are opposed to those of the BAC, has been published in the reports of the BAC. Hence, the BAC and its reports have been an avenue for public expression. Terry Kaan considered the BAC's role in the production of these reports as "instrument of record" to be an important one.[120] John Elliott went further in proposing that the BAC take an additional step in providing specific responses to the views and comments (especially dissenting ones) expressed.[121] Even if the BAC should adopt a different stance on an ethical issue, as inevitably occurs, clear explication of its position and ethical basis would reinforce the BAC's role as an emphatic listener.

CONCLUSION

Since its founding in December 2000, the BAC has contributed to public policy (notably in legislative and regulatory changes) and developed an ethical framework for biomedical research in consultation with its network of consultation parties. The ethical principles that constitute the framework and give shape to ethical practices reflect international norms that define 'good science'. As these principles also reflect local conditions and values, the BAC has been an important mediator or intermediary in the reception of bioethical norms in Singapore.

The BAC has been active in engaging with different segments and levels of society. For some, the BAC has been a source of information, a guide to resolving ethical dilemmas, or a 'watchdog' of some sort. For others, it has been a reason for them to devote time and resources to understanding bioethical issues. And for those whose position on a bioethical issue (such as whether a human embryo has the moral status of 'personhood') is settled, the BAC has been an avenue for expression and also a public agent of record. Public engagement has taken many forms and the above is but a sample of views as to how the work of the BAC could be perceived by the public. This diversity of forms suggests that to think only of the BAC's role as 'public relations',

[119] Interview with Professor Ng Soon Chye, 26 May 2009.
[120] Interview with Associate Professor Terry Kaan, 16 June 2009.
[121] Interview with Associate Professor John Elliott, 27 August 2009.

'feedback collector', 'rubber-stamper' of government pre-determined policies, 'indoctrinator' or the like is too simplistic. The relationship between the BAC and the public is complex, and one that is continuously re-defined. In July 2010, the BAC will be launching a bioethics exhibition (including related activities) developed in collaboration with the Science Centre Singapore and the NUS Centre for Biomedical Ethics. This is a commitment to align the ethical goals of science to the normative expectations of the general public through engagement.

In the initial years of the BAC, there was scepticism about whether bioethics would be a viable long-term enterprise. John Elliott said that such scepticism is unfounded, since the BAC's work will continue to be needed so long as a biomedical research sector is maintained as a national priority.[122] A*STAR Chairman Lim Chuan Poh recently reiterated that the Biomedical Sciences Initiative that was taken up in 2000 is a long-term process.[123] Signs of progress are becoming apparent in the increased output for biomedical sciences from S$6.3 billion in 2000 (3.9% of total manufacturing output) to S$19 billion in 2008 (7.6% of total manufacturing output), more than doubling the number of jobs in research and development (R&D) between 2000 to 2008, and the publication of 1,927 papers in biomedical sciences between 2002 and 2008 by A*STAR's research institutes and filing of 216 primary patents by 2008.[124] The government's commitment to achieve a gross expenditure on R&D of 3.5% of gross domestic product by 2015 is likely to sustain the continued progress of the Biomedical Sciences Initiative.[125] In the light of this, there appears to be more work ahead for the BAC.

---

[122] *Ibid.*

[123] Lim CP. Betting on Biomedical Science: The Nation's Economy Has Evolved Rapidly in Just a Few Decades from Labor-intensive Manufacturing to High-tech Production and Now to Corporate Management and World-class Research, *Issues in Science and Technology*, 22 March 2010.

[124] *Ibid.*

[125] This target is based on the recommendation of the Economic Strategies Committee. See Report of the Economic Strategies Committee, Singapore, *High-Skilled People, Innovative Economy, Distinctive Global City*, February 2010, p. 24.

2

The Impact of the Bioethics Advisory Committee on the Research Community in Singapore

Charmaine KM Chan and Edison T Liu

The Bioethics Advisory Committee (BAC) was established by the Singapore Government in December 2000, shortly after the launch of the Biomedical Sciences Initiative, which is aimed at developing the biomedical sciences into one of the key pillars of the country's economy.[1] The BAC has been tasked to examine the ethical, legal, and social issues arising from life sciences and biomedical research in Singapore, and to recommend policies concerning these issues to the Steering Committee on Life Sciences (SCLS). The establishment of the BAC indicated that Singapore's leaders understood that development in the life sciences cannot advance without a parallel maturation in the ethical handling of research and its outcomes, and therefore aimed to bring clarity, order, and focus in such a way as to reassure both the international research community and the citizenry of Singapore.

How has the BAC contributed to the Biomedical Sciences Initiative thus far? Since its inception, the BAC has released six reports on bioethical issues concerning stem cell research, genetic testing, personal information, donation of human eggs for research, human tissue research, and research involving human subjects. The reports are listed in Annex C at the end of this book. As the BAC celebrates its 10th anniversary this year, we wish to assess the impact that the BAC has had on Singapore's biomedical research community.

[1] For more information about the Biomedical Sciences Initiative, please see: http://www.a-star.edu.sg/AboutASTAR/BiomedicalResearchCouncil/BMSInitiative/tabid/108/Default.aspx.

To this end, we conducted a series of interviews with regulators, researchers, clinician-researchers, and a bioethicist — people who would have encountered the BAC's guidelines in the course of their work. In our conversations, the interviewees shared personal anecdotes of how the BAC's work has affected their own work, gave us their insightful perspectives on the BAC and its functions, and made many suggestions for improvements. Piecing together information from the various interviews, and highlighting the experiences of one of the authors,[2] this chapter examines how the BAC's work in the last ten years has affected the development of the biomedical sciences in Singapore. We then note some challenges facing bioethics in the future and identify some areas for improvement as the BAC continues to mature.

Our overarching view is that the BAC has made a difference. It has done so by being a forum for national consultation around bioethical issues, by being proactive in the choice of questions it asks so as to pre-empt problems, and finally by providing guidelines that carry the Government's imprimatur.

THE BAC'S 'BLUE BOOKS': GUIDING ETHICAL DELIBERATIONS AT THE NATIONAL LEVEL

The BAC has played an important role in leading the deliberation of bioethical issues at the national level. The operational procedure starts off with the selection of a specific issue as the focus of the BAC's consideration, such as stem cell research or data privacy. A working group or sub-committee is convened to study the problem, interview key players, examine international standards, and to draft a consultation paper that includes appropriate recommendations. This document is modified by the BAC as a whole, and subsequently sent to national groups for consultative comments. A revision is made based on the comments received, and the final report, which includes the response documents from the consultative process in its appendices, is sent to the SCLS for discussion and approval. Following confirmation, the final document is published as what is commonly called a 'Blue Book' (for the colour of its cover), representing a set of recommendations that has the imprimatur of the Government at the highest level. Implementation of recommendations is then a matter for regulation or legislative formulation by the relevant Government departments.

[2] Professor Edison T Liu was a member of the BAC from 2003 to 2006.

Most of these recommendations do not require legislative action, but on a national level, the BAC's guidelines have provided important frameworks for regulatory agencies, administrators, and individual scientists for the conduct of biomedical research. Before the establishment of the BAC, there were isolated pockets of bioethical deliberation, but a lack of a coherent set of national guidelines. This meant that different institutions had their own internal guidelines that were adapted from various organisations all over the world.[3] Although many of those interviewed thought that bioethical discussion could still occur without the BAC, there is consensus that the consolidation of a national view on important bioethical issues has been invaluable, just as Eugene Soh, Executive Director of the Research & Development Office of the National Healthcare Group (NHG), said,

> *I believe that the BAC's 'Blue Books' provide deeper insight into ethical issues and help us to better understand the dynamics in research . . . the BAC's guidelines also help to facilitate and support research by providing us with guidance on how best we can think about some of these very difficult issues with regards to human research.*[4]

For individual administrators and regulatory agencies, the complexity of formulating regulatory standards for biomedical innovations and research can be daunting. Being a multi-cultural, multi-racial, and multi-religious society, it has been important for Singapore to have a body like the BAC to consider the various perspectives of the many groups in Singapore to arrive at a national position on bioethical issues. Lee Eng Hin, Executive Director of the Biomedical Research Council, Agency for Science, Technology and Research (A*STAR), thinks that "Bioethics in general is universal, but it also has a national face because different countries will have different perspectives of what they deem as ethical or unethical based on their different cultures and religions".[5] Besides having members of different ethnicities and backgrounds who are representatives of the society, the BAC also consults local interest groups, such as religious bodies, hospitals, scientists, and academic

---

[3] Interview with Professor Lee Eng Hin, 11 September 2009.
[4] Interview with Dr Eugene Fidelis Soh, 7 October 2009.
[5] Interview with Professor Lee Eng Hin, 11 September 2009.

institutions,[6] to ensure that their diverse opinions are taken into consideration. As each group has its own set of ethics, it has not always been easy to produce a single, national perspective on the bioethical issues at hand.

A prime example is the case of embryonic stem cell research. The main ethical issue that the BAC had to address while working on its Stem Cell Report[7] was that of the sanctity of life: Can an embryo be considered as equivalent to a human life? Is it ethically acceptable to destroy an embryo if it may save the life of a human being? Besides seeking input from various religious and professional organisations on its consultation paper, the BAC also conducted a series of public dialogues and gathered feedback from the general public through a focus group discussion and through its website.[8] From its consultation process, the BAC found out that there were several very different opinions on when life began: To Christians, life begins from the point of conception; Muslims believe that 'ensoulment' occurs only four months after conception; and to Taoists, the Spirit enters the body just before birth, when the foetus is fully formed after 300 days. Besides differences in opinions between religions, there were also disagreements on the status of the human embryo even within the same faith. While some people viewed the human embryo as having potential for life and hence were opposed to embryonic stem cell research since it would inevitably mean the destruction of life, others had no problems at all with using embryos for research because they saw the embryos as nothing but clumps of cells. Given that Singapore is a pluralist and secular state, but still taking into consideration the sensitivities of the various religions, the position that the BAC eventually took was somewhere in between: While the embryo enjoys a special moral status, and is therefore deserving of respect, it is not considered to be on equal standing or having the same entitlement as a sentient human person, particularly for embryos less than 14 days old (see below).

Accordingly, in its report published in 2002, the BAC recommended that the derivation and use of human embryonic stem cells be allowed, but only if "there is strong scientific merit in, and potential medical benefit from, such

[6] Available in the annexes of all the BAC's reports is a distribution list stating the organisations/institutions to whom the consultation papers were sent.
[7] Bioethics Advisory Committee, Singapore, *Ethical, Legal and Social Issues in Human Stem Cell Research, Reproduction and Therapeutic Cloning*, June 2002.
[8] *Ibid.* pp. 12 and 13.

research",[9] and under strict regulation by a statutory authority.[10] It was a delicate balance between protecting individual rights and welfare, and respecting individual beliefs, while allowing biomedical research to develop to its full potential for the benefit of Man. The BAC further recommended that derivation of embryonic stem cells from early human embryos is acceptable, provided that the embryos are not more than 14 days old.[11] The BAC considered the 14-day rule to be "an appropriate limit" because that marks the appearance of the primitive streak, which is the time when the central nervous system starts to form.[12] Embryos less than 14 days old would not be considered as sentient beings, hence the derivation of embryonic stem cell from such embryos would not be in violation of the BAC's principle of respect for persons. As the BAC gave a very clear indication of what Singapore's position was on the morality of stem cell research, the field has since progressed rapidly, and Singapore is now a global leader in stem cell research.[13]

THE BAC'S PROACTIVE APPROACH: GUIDELINES THAT ARE LIKE FIRE SAFETY PRECAUTIONS

In describing the BAC's proactive approach in identifying and addressing potential ethical issues before they emerge, Lim Suet Wun, Chief Executive Officer of the NHG, said:[14]

> *It's like fire safety — we don't want to wait until the fire before drawing up the guidelines. In a way, it is a good thing if the guidelines are never used because there has never been a fire (like if someone did something really questionable with embryos, or if patients are harmed). The fact that we don't have fires is positive, though we must remember to remain attentive.*

[9] *Ibid.* p. 30, Recommendation 3.
[10] *Ibid.* p. 33, Recommendation 8.
[11] *Ibid.* p. 29, para 35.
[12] *Ibid.* p. 29, para 34, and p. 30, para 36.
[13] Levine AD. Identifying Under- and Overperforming Countries in Research Related to Human Embryonic Stem Cells, *Cell Stem Cell* **2,** 6 (2008): pp. 521–524. In an assessment on how policies governing human embryonic stem cell research may have influenced the field's development, Singapore was identified as one of the top four overperforming countries in human embryonic stem cell research.
[14] Interview with Dr Lim Suet Wun, 22 September 2009.

Such a proactive stance in addressing issues in advance allows for thoughtful consideration in the absence of high emotions and has had a positive impact on the development of biomedical research in Singapore. As Lim noted, when an incident triggers public outrage, it evokes a "table-thumping, chest-beating, finger-pointing kind of emotive reaction" that can be detrimental to the progress of research.[15] Public confidence and trust in research is difficult to build but very easy to lose. This has been seen in the Alder Hey organ scandal in the UK, where body parts of deceased infants were retained for research without consent. When the official Redfern Report of the Royal Liverpool Children's Inquiry into the scandal was published in 2001, and it was found that that pathologist Dick van Velzen "ordered the unethical and illegal retention of every organ" from all children who underwent postmortem examination during his time at the hospital from 1988 to 1995, there was a great public outcry against the National Health Service.[16] Recovery from the public backlash has been difficult and the research community's reputation suffered immensely. In this case, the fault clearly arose from a lack of clarity about ethical expectations. There is a desire in Singapore to proactively raise issues by a consultative council such as the BAC, to prevent such incidents from happening.

The support of the public is undoubtedly essential for the progress of biomedical research. As Chong Siow Ann, Vice Chairman of the Medical Board (Research) of the Institute of Mental Health, explained,

> *Let's say we are doing a national household survey that we hope will hit 15,000 households. If the general public don't see the value of and don't understand what we are doing, then they will just slam the door on us and we won't get anywhere.*[17]

It is therefore crucial that the public is able to see the potential in proceeding with certain types of research, and remain willing to participate in biomedical research. Because there is often a presumption that the Government has the people's best interest at heart, the public has been mostly supportive of the Government's initiative to advance biomedical research,

---

[15] *Ibid.*

[16] Redfern M, Keeling JW and Powell E, *The Royal Liverpool Children's Inquiry Report*, Liverpool, UK, 2001.

[17] Interview with Associate Professor Chong Siow Ann, 13 October 2009.

and has also been more receptive towards seemingly controversial research than in other countries.

On occasion, the BAC's recommendations, like defined and enforceable laws in any well managed nation, serve to reassure the public of due processes and government oversight. The BAC's guidelines on 'Research Involving Human Subjects'[18] serve precisely this purpose of providing reassurance that biomedical research is carried out in an ethical manner in Singapore. In this 'Blue Book', the BAC provides guidance to local institutional review boards (IRBs) on "the roles and responsibilities of IRBs, researchers and institutions in order to achieve objective and independent ethics review of research proposals involving human subjects",[19] so as to ensure that the safety, welfare, and privacy of research participants are protected. Although not all of the BAC's recommendations eventually become legislation, the recommendations carry weight as national guidelines by virtue of the BAC's relationship with the SCLS. Government agencies would therefore either adopt the guidelines as their own, or would ensure that their regulations are aligned with the BAC recommendations. Prior to the release of the BAC's IRB report in 2004, all research was reviewed either by clinical/hospital ethics committees, or by *ad hoc* committees established within institutions. It was as a consequence of this report, namely the recommendation that all biomedical research involving human subjects in Singapore be "reviewed and approved by a properly constituted IRB before it is allowed to proceed",[20] that standardised IRB processes were formalised on a national level.

BIOETHICS THAT IS 'UNIQUELY SINGAPORE': PRAGMATISM AS THE GUIDING PRINCIPLE

The BAC's recommendations have proven to be very useful as a reference for local agencies when shaping their own guidelines. To aid in its deliberation, the BAC would take into consideration international and other countries' guidelines on the same subject before coming up with its own sets of recommendations that are 'uniquely Singapore'. There is general consensus that the local 'flavour' of the BAC's guidelines is most helpful, because even

[18] Bioethics Advisory Committee, Singapore, *Research Involving Human Subjects: Guidelines for IRBs*, November 2004.
[19] *Ibid.* p. 3, para 10 of Executive Summary.
[20] *Ibid.* p. 4, para 19 of Executive Summary.

though there may be certain ethical principles or concepts that are universally accepted and agreed upon, the local setting, including the social, political, and religious context, plays a very important role in shaping what the society regards as right and wrong.

For the National Healthcare Group, the BAC's recommendations paved the way for the establishment of domain specific review boards (DSRBs), whereby individual hospitals were not required to have their own IRBs but could form 'consortia' with other hospitals to share ethical review boards.[21] In this approach, the hospital cluster consolidated all their IRBs into five DSRBs based on broad but related disease groupings. For example, there is one DSRB that reviews oncology, hematology, pathology, paediatrics, and respiratory medicine protocols for all NHG hospitals, and another that reviews protocols on ophthalmology, psychiatry, neurology/neurosurgery, genetics, and geriatric medicine. This was devised to deal especially with the limited number of expert reviewers for each specialty. Thus, instead of requiring eight cardiac specialists to populate the IRBs of eight hospitals, two can cover the reviews for all eight hospitals. In the absence of a national consensus as defined by the BAC, the hospitals and the Ministries were unlikely to have provided the necessary focus to resolve concerns.

Moreover, these BAC guidelines have also been helpful to the NHG in its application for accreditation by the Association for the Accreditation of Human Research Protection Programs (AAHRPP). One of the AAHRPP standards states that:

> *STANDARD I-3: The Organization's transnational research activities are consistent with the ethical principles set forth in its Human Research Protection Program and meet equivalent levels of participant protection as research conducted in the Organization's principal location while complying with local laws and taking into account cultural context.*[22]

[21] NHG, Bench to Bedside Research Portal for Investigators, DSRB Ethics Review, Singapore. http://www.b2bresearch.nhg.com.sg/Content/content.aspx?id=37d3b0b5-18b0-4310-b616-52996669596c

[22] AAHRPP, AAHRPP Accreditation Standards for Domain I: Organization, Washington, DC, US. http://www.aahrpp.org/www.aspx?PageID=318

The NHG was hence able to use the BAC's 'localised' guidelines as reference to fulfil the stipulated requirement. As noted by Eugene Soh, during their first audit, AAHRPP had checked whether the NHG was in compliance with the BAC's guidelines for research on human subjects, as outlined in its 2004 IRB Report.[23] "If it were not for the BAC, we would not have had a reference source that provided us with guidance on the ethical conduct of biomedical research on human subjects, in the local context."[24]

The BAC's reports have also been a useful complement to the functions of the Health Sciences Authority (HSA). As the national regulatory authority for drugs, health products, and medical devices, the focus of the HSA's clinical trials approval process is mostly on the scientific merit of research proposals. However, while promoting clinical trials, it also has to ensure that the Singaporean population does not unwittingly become exploited in commercial testing.[25] The ethical governance of clinical trials is equally important because they affect whether or not the resulting scientific data will be recognised as legitimate. As such, the HSA has indicated that clinical trials should begin "only when both regulatory and ethics approvals [from the respective IRB/DSRB] have been obtained".[26]

Whilst considering applications, the HSA will also consult its own advisory committee, the Medical Clinical Research Committee (MCRC), whose responsibility is again to safeguard the rights and welfare of human participants in clinical trials.[27] Clinical trials therefore undergo two separate ethics review: one by the respective IRBs/DSRBs, and another by the MCRC. Hence, the BAC's work, in particular its IRB Report, is an integral component in maintaining the credibility of the clinical trials licensing system in Singapore since it provides guidance for the local IRBs/DSRBs. As the BAC's guidelines are in line with international ethical governance standards, it lends credence to the HSA's regulatory system as being of international standing.

[23] Bioethics Advisory Committee, Singapore, *Research Involving Human Subjects: Guidelines for IRBs*, November 2004.

[24] Interview with Dr Eugene Fidelis Soh, 7 October 2009.

[25] Interview with Dr John Lim (Chief Executive Officer, Health Sciences Authority), 5 October 2009.

[26] Health Sciences Authority, Guideline on Application for Clinical Trial Certificate (CTC), Singapore. http://www.hsa.gov.sg/publish/hsaportal/en/health_products_regulation/clinical_trials/guidelines/ctc_application.html

[27] Bioethics Advisory Committee, Singapore, *Research Involving Human Subjects: Guidelines for IRBs*, November 2004, pp. 10–11, para 2.11.

However, given that more clinical trials are moving beyond developed Western countries and into developing Asian countries such as Singapore, it will become increasingly pertinent to have ethical guidelines that are 'uniquely Singapore'. As there are bound to be areas whereby ethical considerations would differ between countries, having 'localised' guidelines like those issued by the BAC as a reference when considering clinical trials proposals have been and will continue to be very useful to the HSA.

> *The norms of the Singapore society are certainly quite different from that of other countries. We are a conservative society, and that is part of our uniqueness. I don't think there is anything wrong with being conservative, and the BAC has struck a fair balance [between conservatism and liberalism] so far. Another feature of Singapore is pragmatism, so the views of the BAC are very pragmatic. The views are not polarised like in other countries such as the US, where there are extreme views on abortion and thus stem cells, so people are always at loggerheads. Again, if you don't address issues proactively before things happen then the situation gets worse. But for us, we are fairly pragmatic: how do we come up with a solution for a problem, and come to some level of agreement? Questions like this reflect the importance of practicality in our discussions. Looking at the reports and statements from the BAC, I think they lean towards being pragmatic. — Lim Suet Wun*[28]

But what does it mean to be 'uniquely Singapore'? It seems that pragmatism is a salient characteristic of Singapore, even when it relates to bioethics. John Elliott believes that it is ethically acceptable for the state to make decisions pragmatically, based on a balance of costs and benefits. Since the citizens' diverse beliefs form the basis of positions from amongst which a Government has to decide, it is reasonable to use the principle of pragmatism to find a resolution that would benefit all, and at the same time avoid the paralysis of irreconcilable positions.[29] As the BAC's recommendations appear to have been neither too conservative nor too liberal, research has been able to move ahead smoothly in Singapore, albeit at the price of some level of increased

[28] Interview with Dr Lim Suet Wun, 22 September 2009.

[29] See Elliott's closing chapter of this book. Also Elliott JM, Ethical Considerations in Human Stem Cell Research, in *Life Sciences: Law and Ethics*, eds. T Kaan and ET Liu, Singapore: Singapore Academy of Law and Bioethics Advisory Committee, 2006, pp. 54–75.

paperwork associated with regulatory requirements. To quote John Lim, Chief Executive Officer of the HSA, "We look for an appropriate ground — not a compromised position, but the appropriate situation for our society, given the time and sensitivities, and norms that we face nowadays."[30]

An example whereby the BAC has devised its own set of guidelines that are 'uniquely Singapore' and on appropriate grounds, given the local context, can be found in its report on the 'Donation of Human Eggs for Research'.[31] Although the BAC's view is generally aligned with the position taken by the UK and US, which is to allow compensation of egg donors for research, the differences are in the details. Here again, the BAC maintained its position, as first stated in the 2002 Human Tissue Report, that tissues donated for research should be outright gifts.[32] However, in view of the fact that women who donate eggs specifically for research will incur a real loss in time and earnings, the BAC agreed that compensation is ethical based on the principle of justice. It recommended that:

> *Egg donors should be compensated only for loss of time and earnings as a result of the procedures required to obtain the eggs, and only if the eggs were obtained specifically for research purposes, and not as a result of clinical treatment. Such compensation should be in addition to any reimbursement of expenses incurred. The relevant regulatory authority should determine the appropriate amount of such compensation.*[33]

The BAC's view on the appropriate amount of compensation is quite distinct from that of the Ethics Committee of the American Society for Reproductive Medicine (ASRM). The ASRM has recommended that all women who donate their eggs, whether for research or fertility treatment purposes, should be compensated for their time, inconvenience, and discomfort. It permits compensation of sums between US$5,000 and US$1000, which the BAC thought could easily pose as an inducement and lead to the exploitation of women who are not that not well-off financially.[34] The BAC had specifically

[30] Interview with Dr John Lim, 5 October 2009.
[31] Bioethics Advisory Committee, *Donation of Human Eggs for Research*, November 2008.
[32] Bioethics Advisory Committee, *Human Tissue Research*, November 2002.
[33] Bioethics Advisory Committee, *Donation of Human Eggs for Research*, November 2008, p. 24, Recommendation 6.
[34] *Ibid.* p. 29, para 4.22.

limited compensation to "loss of time and earnings",[35] which are relatively objective criteria since in practice the quantum would be measured based on the donor's income. For unemployed donors, the BAC has left it up to the relevant authority to decide on the appropriate amount "based on the time spent as a result of the procedures required to obtain the eggs for research".[36] The ASRM's criteria of "inconvenience" and "discomfort" are rather subjective, and a compensation scheme based on these two considerations could prove to be very problematic.

The BAC also disagreed with the position adopted by the UK Human Fertilisation and Embryology Authority that women who donate eggs for research can be entitled to other benefits in the form of treatment services, which could be of unrestricted value. Furthermore, under the Medical Research Council's compensated egg sharing scheme for one of the projects from the North East England Stem Cell Institute, women undergoing fertility treatment who choose to donate some their surplus eggs for the research will have their fees for IVF treatment partially reimbursed.[37] The BAC found such an arrangement to be inappropriate, as the cost of IVF treatment is substantial, and could appear to be a financial inducement, or undue influence, for women planning to undergo IVF treatment to consider egg donation for research.[38] From the BAC's point of view, women who donate their eggs from fertility treatment are not eligible for compensation since they would have undergone voluntary ovarian stimulation and retrieval of eggs in any event.[39] Unlike the other three bodies, therefore, the BAC has recommended that only women who donate their eggs specifically for research should be compensated.

Even though at first glance the BAC's position on the issue of compensation of egg donors for research is consistent with the views in other leading jurisdictions, the BAC has actually issued a set of guidelines that differs in substance. As with its other reports, whilst the BAC ensures that its guidance is aligned with international views, it will also negotiate a position that is appropriate and relevant in the local context.

[35] *Ibid.* p. 24, Recommendation 6.
[36] *Ibid.* p. 23, para 4.32.
[37] Medical Research Council, UK, Press Release, 13 September 2007, *Women Undergoing IVF to Donate Eggs for Stem Cell Research in Return for Reduced Treatment Costs.* http://www.mrc.ac.uk/Newspublications/News/MRC003971.
[38] Bioethics Advisory Committee, *Donation of Human Eggs for Research,* November 2008, p. 21, para 4.24.
[39] *Ibid.* p. 21, para 4.25.

ESTABLISHING SINGAPORE'S INTERNATIONAL REPUTATION AS A BIOMEDICAL RESEARCH CENTRE

One of the most significant effects of the BAC's recommendations has been the assurance, both at the international and national level, that biomedical research is being conducted in an ethical and regulated manner in Singapore. Not only is the BAC comprised of well-respected members of society, it also has a panel of eminent international advisors that provides expert advice on the bioethical issues at hand. Publishing sound reports that are endorsed by such advisors and subsequently accepted by the Government has lent credence to Singapore's reputation as a place with a research framework that is not just scientifically but also ethically sound. As John Lim said,

> *I think it is important for Singapore's reputation that we have a body like the BAC to address all these ethical issues upfront. It is quite foresighted that this had all been set up. It would have been very easy for a fast developing country to just move ahead and promote biomedical research without giving much emphasis to the ethical aspect. The fact that BAC was set up in tandem with the promotion of biomedical research is an indication of how important this aspect is, because credibility [of research data] and also the value of human life is very important in the broader international perspective.*[40]

Room for Improvement

The BAC has clearly contributed much to the successful development of the biomedical research landscape in Singapore thus far. As the Biomedical Sciences Initiative is now in its second phase, whereby its focus is on strengthening capabilities in translational and clinical research, the work of the BAC will become ever more crucial due to increasing translational research involving humans. However, despite its many successes, our interviewees raised areas that could be improved as we move forward. When viewed collectively, the concerns can be divided into three main groupings: first, all recommendations will need to be reviewed and readjusted to emerging societal

[40]Interview with Dr John Lim, 5 October 2009.

trends and attitudes; second, there should be greater connectivity between researchers/scientists and the public; and third, more dialogue between national leadership and the grassroots should be encouraged.

Bioethics as a "Living, Evolving Organism": The Need to Constantly Review Recommendations

> *Bioethics is a living, evolving organism. As society matures, and as research progresses, people's views on certain aspects of research will also change. The BAC therefore needs to review its guidelines every five years or so. It is probably time for the BAC to review its guidelines on human embryonic stem cell research, because the feelings and perceptions towards such research may have changed. — Lee Eng Hin*[41]

In an ever-changing field like bioethics, it is essential for the BAC to constantly reflect on past recommendations, and make revisions when necessary, for Singapore's ethical framework to remain internationally recognised and relevant to the times. As science progresses, society's views change, and ethical standards too will change with time. Given that it has been ten years since the BAC's establishment, and eight years since the BAC's first report, it seems timely for the BAC to review its recommendations, even if it is just to restate its previous position on the matter if still valid. In the words of Lim Suet Wun, "to leave recommendations there and not pay attention to them is to lose the prospective momentum and fall into a reactive mode".[42] Others have similarly indicated that there is a need for the BAC to continually re-examine old ethical issues, to ensure that ethical rigour is maintained.

Although the BAC's recommendations have generally had a positive impact on biomedical research, there have also been some 'misses' whereby research has been unintentionally hampered. As the BAC has been tasked with an advisory role to Government, the BAC's recommendations have mostly been broad and general statements regarding the application of certain bioethical principles in biomedical research. Since the focus of the BAC is not on the details of ethical governance, it avoids issuing recommendations that are too

[41] Interview with Professor Lee Eng Hin, 11 September 2009.
[42] Interview with Dr Lim Suet Wun, 22 September 2009.

prescriptive, thus giving space to the regulators to create their own guidelines. As stated in its IRB Report,

> *We emphasise that it is not the intention of this document to prescribe the specific ethical principles to be applied by IRBs and researchers in the process of ethics governance. We believe that these are professional judgments that are appropriately and properly left to members of IRBs, researchers and other parties involved in the process of ethics governance.*[43]

More Interaction with Scientists and the Public Needed

Some have mentioned that the BAC needs to further its reach to get the lay public more involved in bioethical deliberation. Although the BAC conducts public talks, holds press conferences when its reports are released, and organises occasional public bioethics events, some interviewees felt more could still be done to raise the public profile of the BAC and its work. Another channel through which the BAC gathers feedback is the online discussion forum and e-consultation managed by REACH (Reaching Everyone for Active Citizenry @ Home), but that is limited only to persons who are computer literate. Generally, most of the public and many researchers do not seem to know enough about bioethics, much less about the BAC and what it does.

Many interviewees see the BAC as playing the crucial role of the middle man between the public and the Government on bioethical issues. The BAC is not only an educator of bioethical issues; it is also a mediator of ethical positions. As an educator, the BAC has to distil complex bioethical issues into information that the public can easily understand. In the process of breaking down the ethical issues for the public, the BAC will inevitably be an influence on the public's perception of certain issues. It acts as a moderator of different ethical positions, and by ensuring that discussions are informed and balanced, the BAC can defuse potential misconceptions or prevent extreme moral objections that are without sound basis. The BAC's work therefore aids in keeping the public scientifically open-minded.

[43] Bioethics Advisory Committee, Singapore, *Research Involving Human Subjects: Guidelines for IRBs*, November 2004, para 4.15.

Towards a more Constructive Consultation Process: Promoting a Two-Way Dialogue Rather than a Top-Down Approach

> *The BAC's consultation process is fundamentally pro forma. There should be greater scope for the views of both individuals and groups, to be aired. Instead of people being directed to give their input within a tight time frame, why not give everyone time to think, hold their own meetings to discuss the issues, and so on, before the formal consultation process actually begins. This way, 'grassroots' policy discussion initiatives are encouraged — these play an educative role in themselves, and should serve to inform the BAC's deliberations more helpfully than they have to date.*
>
> — *Justin Burley*[44]

Many have expressed the view that the BAC's approach in gathering public feedback on its consultation papers has been too top-down in nature, and that improvements are necessary so that ethical discussions can be more inclusive and engaging. The standard protocol for the BAC's public consultation process is to send out a consultation paper, which will already identify the salient ethical issues and contain the BAC's views on the subject, to target religious and professional organisations, for their comments and feedback. Since answers to the questions have already been brought to the table, there can be little room for truly meaningful and thought-provoking ethical deliberation. As Eugene Soh explained,

> *This is more like a 'buy-in' process rather than a consultation process. What you really want is a dialogue to happen before the guidelines are even drafted. You still need something to spurn the discussion because you cannot just start with a blank piece of paper. But in terms of formulating the guidance, you need to be able to go there with questions, and not answers. It ought to be a dialogue process, and not, "What do you think of my answers?" You are not asking people to buy-in to the way you think ... It takes more time to do a dialogue-type of consultation, but the product that comes out of it will have the pulse of the ground.*[45]

[44] Interview with Associate Professor Justine Burley, 15 September 2009.
[45] Interview with Dr Eugene Fidelis Soh, 7 October 2009.

This view was shared by Chong Siow Ann:

> *There is a need to consult at the ground level, to find out what are the needs and problem areas that need to be addressed. This has to be from the very beginning, not after you have already formulated the guidelines.... You will be rewarded with better feedback if you make a better effort to engage the ground.*[46]

It is felt that the BAC's approach comes across as being merely procedural because appreciation for the feedback that it receives seems to be lacking on the part of the BAC. This perception — that the BAC does not take into serious consideration the opinions and comments it receives from the public — seems to stem from a missing sense of 'reciprocity'. As the BAC does not always make a formal response in its report to address issues raised by the local interest groups in their feedback, the consultation process appears to be one-directional and offers little incentive for groups to respond. There is a real need for the BAC to make a greater effort to communicate with the public, and maintain a more constant public engagement, so as to keep the public, or even researchers for that matter, interested in being part of the bioethical deliberations.

CONCLUSION: A BALANCING ACT

> *The BAC's track record is already solid. Its output is, on the whole, well reasoned. That is, the content of the Reports the BAC has issued reflects key aspects of Singaporean culture and society without departing from universal norms that are publicly justifiable in an international medical research context... That said, the BAC's output also rather lags behind developments in the biosciences. It is perhaps time for the BAC to think more seriously about how research is moving forward. In this vein, it can be noted that much of the research that is being conducted in Singapore and elsewhere world is integrative: it is informed by ideas, technology, skills and techniques from different disciplines. This suggests that a broader approach to bioethics is indicated. — Justin Burley*[47]

[46] Interview with Associate Professor Chong Siow Ann, 13 October 2009.
[47] Interview with Associate Professor Justine Burley, 15 September 2009.

While science progresses with remarkable speed, ethics often lags behind. Yet, the challenge is not to be prematurely proscriptive in a changing field. There are many new rapidly developing fields of research that beckon: nanotechnology, synthetic biology, personal genome sequencing, and tissue engineering — just to name the top few contemporary topics which deserve the attention of the BAC.

Another delicate balancing act for the BAC relates to its role as the national advisory body. Should the BAC remain more general in its recommendations, or should the BAC be more proscriptive in its guidelines so as to facilitate the regulatory process? As mentioned earlier, the BAC's intention in its reports is to establish broad guiding principles for general guidance, and regulatory details are left to the discretion of relevant authorities. However, the lack of clarity on how these broad ethical principles are to be applied, or implemented for regulation purposes, could slow down the entire regulatory process. In a 'uniquely Singapore' way, there is always an emphasis on ensuring that things run smoothly and efficiently, so there is an unspoken expectation that the BAC would be more definitive in resolving controversies. But Eugene Soh wishes to err on being less proscriptive:

> *I think that the BAC's leadership lies in keeping to the ethical principles and providing guidance on how best to think about difficult ethical issues. The BAC needs to be more careful because their guidance is provided at the national level. There are many different types of hospitals and research institutes, so there are many different contexts to consider when it relates to regulation. If you want to go down to the operational decision level, that is a question of corporate governance, and not ethical governance. So don't corrupt ethical governance with corporate governance because that is where you run the risk of stepping on issues that could otherwise have facilitated the development of better ethical governance ... We are talking about think tank/thought leadership here, not corporate leadership.*[48]

The reality is that with each interface between technology and humanity, issues of ethics will undoubtedly arise. The BAC, or an organisation of the

[48] Interview with Dr Eugene Fidelis Soh, 7 October 2009.

same character, will need to be in place for societal convergence to take place. How the Bioethics Advisory Committee evolves in the future will depend on the nature of the Singaporean society as we mature as a nation and as a global player in biomedical innovation.

3

Engaging the Public: The Role of the Media

Chang Ai-Lien and Judith Tan

Any attempt at science communication or public engagement does not take place in a vacuum. It happens in a world saturated with news headlines and pervasive cultural images. The mass media can be a powerful force in influencing how people talk about science, scientists, and scientific evidence.[1] Often, the messages interpreted by the public may not necessarily be those intended by the producers — in this case the researchers. How people respond may be influenced by class, gender, sexual and ethnic identity, as well as a wider cultural context. Experimental work and statistical analysis of trends suggest that the media can set the agenda around what problems society is facing and how we should be setting priorities. The mass media is clearly able to engage its audience to some extent, providing space for exploring dilemmas or 'what ifs', and raising questions about the potential social consequences of science.

Singapore's local media, the *Straits Times* (ST) in particular, has played a major role to inform and engage the public. ST — at 165 years old in 2010 — is the oldest existing newspaper in East Asia. It is Singapore's paper of record, and the main English language newspaper with an established track record and a daily readership of over 1.4 million people. Its readers are relatively affluent and well-educated. Forty-three per cent of its readers are occupationally classified as professionals, managers, executives, and businessmen with relatively high spending power. While media ownership in Singapore is carefully regulated and media content monitored, foreign publications, cable

[1] Kitzinger J. The Role of Media in Public Engagement, in *Engaging Science: Thoughts, Deeds, Analysis and Action*, ed. J Turney. London: Wellcome Trust, 2007, pp. 45–49.

television, the British Broadcasting Corporation (BBC) radio, a range of Internet sites, magazines, and books make up a substantial fraction of information provision.[2] Most international broadsheets and magazines such as the *International Herald Tribune, Far Eastern Economic Review, Financial Times, USA Today,* and *Newsweek* are available. Foreign language newspapers such as the *Frankfurter Allgemeine Zeitung* and newspapers from Japan and India are also available.[3] As Singapore pursued its biomedical drive, ST has chronicled the Republic's journey from its first steps in the latter half of the 1990s into the complex and evolving world of research and bioethics.[4]

ROLLING OUT SINGAPORE'S BIOMEDICAL SCIENCES INITIATIVE: A MULTIBILLION-DOLLAR GAMBLE

The Biomedical Sciences Initiative (BMSI) was a 'blue-print' for Singapore to sustain its economic growth through the knowledge industries. It was drawn up in an all-night session by Mr Philip Yeo, then Chairman of Singapore's Agency for Science, Technology and Research (A*STAR), with three of Singapore's top doctors, Tan Chorh Chuan and oncologists John Wong and Kong Hwai Loong.[5] The Government gave the green light and the plan was announced in 2000 — one day before the unveiling of the Human Genome Project. Nearly US$2 billion was devoted over the next five years to the development of public and private sector biomedical research, aiming to provide by 2007 at least 4,000 jobs in the industry.[6]

To ensure it succeeded in its research push, Singapore needed to maintain a critical mass of scientists, and ensure a research-friendly environment, acceptable to both the scientists and society. While there was no lack of activity in schools in the 1980s and 1990s to excite young minds about science, it certainly did not appear to have led to careers in research and development (R&D). A 1993 survey showed some 15% of the 6,700 research scientists

[2]Gomez J. Freedom of Expression and the Media — part of a series of baseline studies on seven Southeast Asian countries, *Article 19*, December 2005, London. Available at http://www.article19.org/pdfs/publications/singapore-baseline-study.pdf.
[3]*Ibid.*
[4]Chang AL and Soh N. Ethics Trail Medical Research, *The Straits Times*, 30 March 2001, p. H18.
[5]Van Epps HL. Singapore's Multibillion Dollar Gamble, *Journal of Experimental Medicine* **203**, 5 (2006): 1139–1142.
[6]Chia A. Biomedical Hub Will Provide At Least 4,000 Jobs: Philip Yeo, *The Business Times*, 11 May 2002, p. 2.

were foreigners. More than half of the researchers in the NUS science faculty were also from overseas.[7]

One possible reason for the low interest in R&D as a career choice among local students in the late 1980s and early 1990s was that many of the programmes and activities were meant to nurture interest in science and make learning fun. Since most of these activities were not directly related to school grades, parents and even some principals did not encourage older students to spend too much time on them. Instead, more pre-university students opted for business and accountancy and other university courses, even though they did science up to pre-university (or high school equivalent).

A survey conducted on three batches of 152 science and chemistry secondary school teachers attending training courses at Singapore Polytechnic between 1997 and 1999 asked whether bioethical issues were being taught in schools even though they were not in the curriculum. The results showed that bioethics was the topic least taught in Singapore.[8] The survey also found that over 90% of respondents were in favour of bioethics education. There followed a trend to include current bioscience and bioethics issues in the education syllabus in schools and tertiary institutions. Such instruction is vital in preventing prejudice in an increasingly complex and potentially dangerous environment in ethical decision-making.[9]

Then Trade and Industry Minister George Yeo committed resources and reorientated the entire education system in 2000 to achieve Singapore's long-term development in the life sciences as an important pillar of the economy.[10] Singapore immediately set out to lure some of the world's best scientists here, with the idea that they would help make the Republic a more enticing option for the best researchers and science students needed to get the science machinery up and running.[11] Mr Yeo's aggressive recruitment efforts resulted in Singapore becoming home to a number of acclaimed scientists. Among

---

[7] Seah L. When I Grow Up, I Don't Want To Be A Scientist. Where Have All The Young Researchers Gone? *The Straits Times*, 2 November 1995, pp. L1–L2.

[8] Macer D and Chin CO. Bioethics Education Among Singapore High School Science Teachers, *Eubios Journal of Asian and International Bioethics* **9** (1999): 138–144.

[9] *Ibid.*

[10] Khalik S. Setting Standards for Ethics of Life-sciences Research, *The Straits Times*, 9 December 2000, p. H6.

[11] Van Epps HL. Singapore's Multibillion Dollar Gamble, *Journal of Experimental Medicine* **203**, 5 (2006): 1139–1142.

them are Sir David Lane, discoverer of the p53 tumour suppressor gene; Edison Liu, former head of the Division of Clinical Sciences at the National Cancer Institute in the US; Japanese cancer researcher Yoshiaki Ito; molecular biologist Alan Colman; and Nobel laureate Sydney Brenner.[12]

To ramp up R&D activities locally, A*STAR, a key government agency responsible for developing Singapore's biomedical research capabilities, offered a variety of awards and scholarships to scientific talent at all stages of their careers — from young students and undergraduates to post-doctoral fellows. These included a joint initiative by A*STAR and the Education Ministry to match secondary school students with a researcher in an A*STAR Research Institute during the year-end school break, and the National Science Scholarship (MBBS-PhD) for students aspiring to be clinician-scientists.

Since this biomedical drive was launched ten years ago, billions of dollars have been pumped into the sector to train and attract scientists, build up facilities, and attract large biomedical companies here.[13] This investment commitment has not been hit by the economic crisis, and the Government has allocated S$13 billion for public research spending for the period 2006–2010.[14] Given an output of $20.7 billion in 2009, the Government seems confident that the thriving biomedical science sector will be able to hit its target of $25 billion by 2015. There are currently more than 100 biomedical companies in Singapore employing about 13,000 people. Around 50 firms are involved in research and development and 40 in manufacturing, while 30 companies have their regional headquarters here.[15]

BAC'S ACTIVITIES AND NEWS COVERAGE

The authorities recognised from the outset the importance of doing not just good science, but ethically sound research that could stand up to international scrutiny. In an era where our economy was more and more dependent on science, and one where the stuff of science fiction was being transformed into reality, it was imperative that the public be informed and involved. This was

[12] *Ibid.*
[13] Loo D. Research Drive On The Right Track, Says Tharman, *The Straits Times*, 16 February 2007, p. 3.
[14] Johnson P. *Intellectual Capital for Communities in the Knowledge Economy*. Singapore: National Library Board, 2008.
[15] Vaughn V. Biomed Sector To Grow By Up To 10%, *The Straits Times*, 18 March 2010, p. B18.

particularly important in the more controversial areas of study. Singapore's Bioethics Advisory Committee (BAC) has played a critical role in building awareness within broader society — from scratch — of the new areas of research being embarked on. Working together with the media and interested parties, the committee has highlighted the issues involved and how they affect the man on the street.

It was clear early on that biomedical research had immense potential to improve medical treatments and quality of life, and was set to create jobs here and boost the economy in the long run. Yet, however ethical the research might be and however unrealistic any public fears might be, the effort would come to naught if people mistrusted the research and rejected it. Consequently, it was felt that more had to be done to make the public more a partner instead of an adversary as Singapore strove to create a research-friendly environment. After all, members of the public were also needed as active participants in donating tissue and volunteering for trials. The first step was to decide what could and could not be done here. So the BAC was set up in 2000 to look at legal, ethical, and social issues arising from biomedical research and its applications, identify the concerns, and develop recommendations.

Singapore was among the first countries in the world to engage its public in a focused national effort to identify issues in the biomedical sciences as a social concern. Before this, the Republic had no specific guidelines for contentious areas such as human cloning. While it was able to learn from countries such as Britain, Australia, and the US in shaping its approach to bioethics, guidelines from other countries could not be adopted wholesale. The island state had its own set of challenges to deal with. The country is a melting pot of races, cultures, and religions. Close to 5 million people — Chinese, Malays, Indians, Eurasians; Buddhists, Muslims, Christians, Taoists, Hindus — live alongside one another. In setting out a clear framework for what is and is not allowed in the burgeoning field of biomedical research, the BAC needed to help individuals contend with controversial issues.

The idea then was to increase public awareness of the research and its implications, so that contentious issues could be discussed and addressed, without creating social rifts. Setting up the BAC was timely, because in 2000, apart from having no specific guidelines on controversial research such as that involving human embryonic stem cells and some genetics research,

Singaporeans in general had little knowledge of the type of research being embarked on here, or its applications.[16]

In the past decade, there have been over 220 reports on the BAC in the English print media under the Singapore Press Holdings (SPH) umbrella, which includes ST, *The Business Times* and *The New Paper*. Of these, 152 reports were in ST alone. These were strictly news reports and features; the total excludes letters from the public. In fact, ST has tracked every BAC announcement and public consultation exercise, and continues to do so. Over the same period, there were 93 reports in the Chinese print media under SPH and 34 in the Malay language press.

THE GENETIC MODIFICATION ADVISORY COMMITTEE (GMAC)

The Genetic Modification Advisory Committee (GMAC) and the BAC have been the two key players leading the effort to increase public education in science and bioethics in Singapore.

Just a year before the BAC was formed, the Government set up GMAC as a watchdog group for genetically modified foods. In line with efforts to increase public understanding in new and contentious areas of science, it was formed to look at genetically modified foods and products and ensure their safety; oversee and give advice on the research, production, handling, and testing of genetically modified organisms and products made from them. It would also work on increasing public understanding and allaying any fears.[17]

Unlike their European counterparts who protest against what they call 'Frankenstein food', the majority of Singaporeans do not appear to be particularly concerned about the issue and trust the Government to ensure their food is safe to eat.[18] According to a 2000 survey by the Asian Food Information Centre, a non-profit organisation which provides information on food safety and nutrition, less than 2% of people in Singapore know what a genetically modified organism (GMO) is. People were not aware that such plants and animals incorporate genes to give them special properties, and that some GMO

[16]Chang AL. Unpublished interviews, 2000.
[17]Chang AL. On The Harvest Bandwagon, *The Straits Times*, 6 June 1999, p. 36.
[18]Tan J. Interview with Professor Simon Tay, Chairman, Singapore Institute of International Affairs, 6 January 2010.

products are already on supermarket shelves here. According to the same survey, only two out of 120 people in six focus group discussions here were able to explain what genetic engineering is, but most were quite accepting of the technology when it was explained to them.[19]

Like the BAC, GMAC has set up a comprehensive website where people can get information on genetically modified products and the technology entailed. It provides scientific advice to the Agri-Food and Veterinary Authority (AVA), which is responsible for putting in place standards that are in line with international guidelines for genetically modified foods. The AVA carries out independent assessments of applications from companies intending to market genetically modified foods here, based on recommendations of the Codex Alimentarius Commission, a body under the World Health Organization and the Food and Agriculture Organization. In 2003, the commission established an internationally recognised set of guidelines prescribing how genetically modified foods are to be assessed for potential toxicity, nutritional equivalence with conventional counterparts, and unintended effects that could result from the insertion of foreign genes.[20]

GMAC also came up with a set of guidelines in 2006 for scientists here, based on best practices overseas.[21] Under the guidelines, organisations carrying out research on GMOs need to have their research proposals assessed by an institution's biosafety committee. Those conducting experiments with the potential to infect a significant part of the population or harm the environment — such as experiments that involve the avian flu virus — will have to be approved by GMAC.[22]

There is growing public awareness, however. In May 2005, GMAC commissioned a nationwide survey to better understand Singaporeans' knowledge, attitudes, and perceptions of genetic modification technology. This survey, conducted by NUS Consulting, was a follow-up of a similar study conducted four years ago in May 2001. Results of the data collected through interviews

[19]Author unknown. Few People Here Aware of GM Products, *The Straits Times*, 23 November 2000, p. H13.
[20]Droge P, Tan KP, Qiu SJ and Uttam S. Seeding A Second Green Revolution, *The Straits Times*, 2 January 2010, p. B17; Goh SY. All Food Here, Including GM Food, Safe to Eat: AVA, *The Straits Times*, 25 March 2008, p. H6.
[21]Chen HF. New Guidelines for Research Involving Genetic Modification, *The Business Times*, 19 May 2006, p. 14.
[22]*Ibid.*

with 600 Singaporean adults at public places found that 40% of Singaporeans had heard of the term 'genetic modification', and that those who had were more informed and held fewer misconceptions about the subject matter as compared to four years ago. Among those that had heard of the term 'genetic modification', attitudes towards genetically modified foods were favourable. More than two-thirds believed the technology would increase food production and confer benefits to the farmers. Just as many would be willing to buy genetically modified foods if they offered tangible benefits such as better appearance, lower price, or improved taste. Most of those interviewed reported learning about genetically modified foods from the mass media such as newspapers, TV, magazines, and radio.[23] In fact, over 90% of consumers receive information about food and biotechnology primarily through the popular mass media. Both the BAC and GMAC appear to have worked in similar ways to make people in Singapore more aware of scientific activities and the potential ethical problems they may pose, with slow but tangible results.

PUBLIC AWARENESS

Scientists and researchers — here and worldwide — worry that public misunderstanding of science could impede scientific progress. An implicit assumption is that a scientifically educated public will be more open to scientific research.[24]

There are limitations to the extent to which public engagement is useful in determining research priorities and how research is to be carried out. It has been suggested that science as a technical field could not be regarded as a "democratic activity".[25] But demands for greater public engagement should not be rejected, especially since important scientific research is to a large degree sponsored by the public.

However, where public values are allowed to impinge directly on science and research, results are not always beneficial. Take, for instance, research

[23] Genetic Modification Advisory Committee, Survey Indicates Singaporeans' Knowledge and Attitudes Towards Genetic Modification Have Improved Slightly Since 2001, 25 January 2007.
[24] M Nisbet. Who's Getting It Right and Who's Getting It Wrong in the Debate About Science Literacy? Opinions Clash Over the Best Way to Bolster Public Support for Science, *Science and the Media*, 9 June 2003. Available at http://www.csicop.org /specialarticles/show/whos_getting_it_right_and_whos_getting_it_wrong_in_ the_debate_about_science/.
[25] Taverne D. Let's Be Sensible About Public Participation, *Nature* **432**, 7015 (2004): 271.

using embryonic stem cells.[26] France, Germany, Russia,[27] and other countries that outlaw such research for essentially religious reasons are deliberately choosing to cut off work that could have developed into important therapeutic applications across a range of medical conditions.[28] Although this is a legitimate choice given the value systems involved, the challenge facing policy-makers here is not to find ways of dismissing such positions. Rather, it is to develop ways by which scientific research could be carried out within particular value systems.[29] In other words, since science is not value-free, the issue becomes which values are to be the guiding ones that motivate public interest.

In Singapore, the BAC undertook extensive local consultation with experts, interests groups, and the public before it came up with guidelines on human embryonic stem cell research and other bio-science issues. A draft document was sent to more than 30 religious, medical, scientific, and patient-focused groups for feedback; after which the committee held dialogues with them.[30] With the help of the Feedback Unit of the Ministry of Community Development, Youth and Sports, a focus group discussion was also held with 39 participants from different walks of life. There were several press briefings to receive views from the media and to keep the public informed. Members of the public were also invited to provide their views via e-mail, and the BAC received many such messages daily. The whole process took ten months of consultation with the public and various religious and professional groups.

Issues relating to stem cells are difficult as they revolve around the question of whether an embryo is a living human, and whether killing it for the stem cells is tantamount to taking a life. The moral issue here was not taken in isolation and had to be weighed against the consideration that lives can be saved by using stem cells.[31] As the BAC pondered over its recommendations,

---

[26] Dickson D. The Need to Increase Public Engagement in Science, *Nature* **432**, 7018 (2004): 271.

[27] Wheat K and Matthews K. World Human Cloning Policies, *Innovative e-Learning in Reparative Medicine* (NOVAe-MED). Available at http://novae-med.bg-id.com/base1/5.pdf.

[28] Dickson D. The Need to Increase Public Engagement in Science, *Nature* **432**, 7018 (2004): 271.

[29] *Ibid.*

[30] McCoy P. Govt Biomedical Watchdog Body May Be Set Up, *The Straits Times*, 18 November 2001, p. 5.

[31] Chuang PM. Biomed Ethics Guide: Open Approach, *The Business Times*, 13 September 2001, p. 14.

Minister Yeo wrote that as the world watched, Singapore "should not adopt an amoral approach" but, at the same time, it "should also not be doctrinaire . . . [that] the potential for good, for curing disease and saving lives . . . [is] waved aside in the name of dogma".[32] Public feedback also flowed in — most of it concerning the use of human stem cells in biomedical R&D.

The BAC took the position that while the embryo deserved respect and had a moral status as a potential human being, this status was 'different' from that of a baby or adult. It drew praise from scientists and observers around the world for being 'sensible', 'sound', and 'well thought-out'; but not all religious communities welcomed the committee's conclusions.[33] The National Council of Churches of Singapore said that while the issues and regulations were generally sensitive and sound, it was unacceptable to abort the embryo or foetus for research. This was echoed by the Catholic Medical Guild. Made up of about 300 medical doctors, the Guild was disappointed that the BAC had chosen to condone and even promote research that involves what it termed as 'the killing of human life'. Its representative, Dr John Hui, said once this was set in motion, it would "draw us down a spiral of moral deconscientisation that we may never recover from".[34]

It was based on the BAC's recommendations that a comprehensive legislative framework was set up by the Health Ministry. It came up with guidelines for the licensing, control, and monitoring of all human stem cell and cloning research conducted in Singapore. This was to ensure all such research work carried out in Singapore would be ethically sound and carefully monitored. Safeguards were also put in place to prevent the sale and other commercial dealings in human tissue for the purpose of biomedical research.[35]

To gain public trust, scientists look towards involving the public in their work, turning first to education and creating awareness as the answer. In the early 1970s, annual surveys were conducted by the US National Science

[32]Yeo G (Minister for Trade and Industry). Don't Let Dogma Block Potential of Stem-Cell Study, Speech at the 35th annual dinner of the Institution of Engineers Singapore, 17 September 2001; reported in *The Straits Times,* 19 September 2001, p. 18.
[33]Soh N and Chang AL. Panel Sought Various Views for Guidelines, *The Straits Times,* 22 June 2002, p. H19.
[34]*Ibid.*
[35]S Balaji. A Legislative Framework for Stem Cell Research in Singapore, *SMA News* **35**, 12 (2003): 1 and 4.

Foundation to determine the levels of public understanding.[36] The phrase 'public engagement' can be simply a way of repeating the straightforward goal of educating lay people about scientific facts. Sometimes it refers to a wish to inform consumers about the value of peer review or, on the other hand, to remind them that scientific findings are always unplanned or unforeseen.[37] Other times, it is used to describe activities designed to inspire youngsters with the excitement of science, in the hope of recruiting the scientists of the future. But in this context, 'public engagement' is used to imply the wish to consult citizens or involve them in setting the R&D agenda and reflecting on the social context and consequences of diverse choices. Public engagement with science is not so much a new label as a new concept as we move from 'listen and learn' to engagement.[38]

Yet, as a term, it means different things to different people. For some, it means dialogue, where there is genuine discussion between scientists and the public; for others, it is about the importance of the public voice being fed into scientific policy-making; for others still, it covers the full array of activities in which scientifically trained or active individuals interact with lay people without any scientific background.[39]

As Singapore moves towards a society that relies more on research and innovation to drive the country's economic growth, issues have emerged that were likely to become areas of public concern. These include genomics and stem cell research, being areas that have given and are continuing to give rise to distinct sets of ethical and social dilemmas.[40] When conducting genetic testing, for instance, it is important to observe the welfare, safety, and religious and cultural perspectives and traditions of individuals. In this aspect, the extent of public awareness could be seen as either a person who is knowledgeable or one who is merely interested. The one who has the knowledge

[36]Wynne B. Public Engagement as a Means of Restoring Public Trust in Science — Hitting the Notes, but Missing the Music? *Community Genetics* **9**, 3 (2006): 211–220.
[37]Kitzinger J. The Role of Media in Public Engagement, in *Engaging Science: Thoughts, Deeds, Analysis and Action*, ed. J Turney. London: Wellcome Trust, 2007, pp. 45–49.
[38]Boon T. A Historical Perspective on Science Engagement, in *Engaging Science: Thoughts, Deeds, Analysis and Action*, ed. J Turney. London: Wellcome Trust, 2007, pp. 8–13.
[39]Matterson C. Engaging Science: Creative Enterprise or Controlled Endeavour? in *Engaging Science: Thoughts, Deeds, Analysis and Action*, ed. J Turney. London: Wellcome Trust, 2007, pp. 8–13.
[40]Wilsdon J and Willis R. *See through Science: Why Public Engagement Needs to Move Upstream*. London: Demos, 2004.

often has an awareness of what the processes are, while a person who is interested may just read the newspaper or talk to friends in order to learn more and to keep abreast of changes and developments.[41] It is often the former who engages actively in dialogues and discussions.

Being engaged is often a strong predictor of whether the individual will vote or participate in meaningful discussions on issues involving the community at large. And with such controversies over issues such as genetically modified foods, for example, people should be questioning scientists and researchers more. In spite of its benefits, there is as much opposition to the science as there is support and the public has already shown how it can be a powerful force — pulling the plug on some very promising research, and showing the scientific community strong public engagement is a force to be reckoned with.[42] One glaring example is how British and European consumers rejected genetically modified crops because of bad press and because companies developing such technology had not been transparent enough as they did not see the need to explain their work. Despite the lack of evidence of genetically modified foods being a health hazard, and the considerable potential of these 'super-crops' to alleviate the world's food woes, they were widely rejected.

TRUST

The Singaporean public generally recognises the value of science and technology but, at the same time, does not adequately understand the issues related to or arising from them. Its general attitude is that of trust in government policies, and it has remained so even in the research arena. Response to consultation efforts, therefore, has remained low-key.[43]

It is not enough for the public to have scientific literacy. It must also be 'attentive' in following the progress of scientific issues on a regular, perhaps even daily, basis.[44] Attitude towards science and technology is another

[41] Milbrath LW and Goel ML. *Political Participation.* Chicago: Rand McNally, 1977; and Finkel SE. *Causal Analysis with Panel Data.* Thousand Oaks, CA: Sage, 1985.
[42] Chang AL. Speak Up on Science Issues Before It's Too Late, *The Straits Times,* 11 August 2009, p. A2.
[43] Goh CL. Singapore's Quiet Lobbyists, *The Straits Times,* 28 October 2006, pp. S4–S5.
[44] Organisation for Economic Co-operation and Development. *Science and Technology in the Public Eye.* Paris: Organisation for Economic Co-operation and Development, 1997, pp. 9 and 11.

relevant dimension of public understanding. When people obtain information of a scientific or technical sort, they generally tend to have either a favourable or an unfavourable bias towards that information. Since most adults here have relatively little direct experience with science and technology issues, it is likely that many people have relatively weak attitudes towards the subject. In Singapore, the regulatory system is not as transparent in decision-making as in certain Western European countries. Yet despite that, there is a lot of public trust.[45] Take for instance the hygiene grading that the National Environment Agency (NEA) gives to food retail outlets such as restaurants, cafes, snack bars, supermarkets, mobile food wagons, and food caterers. Retail food establishments are given a grade by NEA based on the overall hygiene, cleanliness, and housekeeping standards of the premises, but how the places are graded is not transparent. Yet, the process is accepted without question by the public.[46] Public trust here may be defined as the responsibility the public places on the government to care for their interests. But because there is in some areas hardly any or no transparency, it is similar to the trust consumers put in brands. For example, Mercedes-Benz is trusted to give luxury quality.[47]

As journalists tracked the biomedical drive over the last decade, interviews with 'the man on the street' showed growing awareness of the type of research being done here and the ethical issues involved, particularly among the young. For example, half of those interviewed in a random poll had heard of terms like stem cells and gene therapy, although fewer than one in ten could explain correctly what they meant. Despite extensive news coverage and the BAC's outreach efforts, including public workshops and forums, most people in Singapore remain uninvolved and uninterested in developments in the biomedical sphere.

TOPICS THAT DREW THE MOST RESPONSE

As is the case elsewhere, topics here which were most talked about were those that affected people personally, and those that went against their religious views.

[45]Tan J. Interview with Professor Simon Tay, Chairman, Singapore Institute of International Affairs, 6 January 2010.
[46]*Ibid.*
[47]*Ibid.*

Embryonic Stem Cell Research and Therapeutic Cloning

In 2002, the BAC came up with its first set of recommendations, in one of the most controversial areas that was widely debated overseas with vastly different results: the use of stem cells in research, as well as therapeutic and reproductive cloning. Among its recommendations were (subject to certain ethical requirements) that embryonic stem cell research be given the go-ahead for embryos less than 14 days old, and that such cells could be taken from aborted foetuses, surplus embryos from fertility treatment, or embryos created through IVF and other techniques.[48] It also approved, under extremely stringent conditions, the cloning of embryos for producing stem cells. Cloning of humans was given an outright ban. In coming up with its recommendations, the committee went through (as noted earlier) an exhaustive ten-month consultation process with the public. It also commissioned papers from scientists and doctors.

Of the main religious groups here, most indicated their acceptance. The Christians, in particular the Catholics and their local medical guild, came out strongly against such research, because of their belief that life began at conception. At the time, there were also several letters to ST's Forum page which echoed these views and focused on the rights of days-old embryos as human beings, as opposed to cells to be harvested for research. Most of the people who attended discussions on the issue, however, seemed to understand the need and value of embryonic stem cell research, and were not against it. The majority accepted 14 days as the cut-off point for using embryos, although fewer agreed that embryos should be cloned and used for research. The BAC's final guidelines took into account all views, while acknowledging that it would be impossible to satisfy everyone. They were praised internationally for being sound, and were accepted by the Singapore Government, which prohibited reproductive cloning by law in 2004. Stiff penalties were put in place. Those who broke the law, for example by attempting to clone a human, faced up to ten years in jail and a $100,000 fine.[49] Again, public reaction to these developments was positive. During this consultation process, some wrote

[48] Low E. Set Up Body to Regulate Stem Cell Research: Panel, *The Business Times,* 8 January 2002, p. 7.

[49] Chang AL. No to Human Cloning But Yes to Some Stem Cell Work, *The Straits Times,* 3 September 2004, p. 3.

in to express support while others asked for clarification. Few, if any, had objections.

Using Personal Information in Research

In Singapore, genetic testing has become part of clinical medicine and its role will continue to grow. There are many genetic tests available on a clinical basis and these are mainly initiated by the doctor when the medical complaint has a possibly genetic origin. A small proportion is patient-driven and this tends to be in the area of carrier, pre-natal, and pre-symptomatic diagnosis which can help predict whether they, or their children, will fall prey to a particular disease. While some tests, such as those for Down's syndrome in the unborn baby and blood disorders such as thalassaemia have been available for years, newer ones include those for a host of rare cancers caused by a single gene gone wrong. In 2005, the Singapore Government accepted 22 recommendations on genetic testing and research made by the BAC. While testing for genetic disorders is permitted, any test aiming to alter a gene is not allowed, even if it seeks to get rid of a disorder.[50]

Owing to the sensitive nature of genetic information derived from genetic testing and the impact it can have on the privacy of the individual and his or her family, the BAC indicated that it is important that such testing be conducted responsibly and ethically. So getting a sample tissue from a person for genetic testing without his or her knowledge is considered non-consensual or deceitful. At the same time, the BAC also recommended that no tissue or DNA should be traded on a commercial basis in Singapore.

Other areas which drew greater than average public feedback were those that were seen to have a direct impact on people's lives. Before the BAC's recommendations on personal information in 2007, there were no existing laws to protect personal information in biomedical research, unlike, say, banking or medical records. And the worry, based on public feedback, was that such information would land in the wrong hands, such as those of an insurer who could increase premiums based on susceptibility to illness. In 2006, the BAC made the assurance that no personal information generated from biomedical research would be available to third parties — not even to employers and

[50]Chen HF. No Coercion Allowed in Genetic Testing: BAC, *The Business Times*, 26 November 2005, p. 0.

insurers who may be in a position to use such data to discriminate against prospective employees or clients.[51]

The BAC said it was understandable that those who participate in biomedical research were concerned they might lose out in terms of their privacy, autonomy, and personal interest — something that needed to be addressed to make sure the community and the public would support biomedical research and, more importantly, participate in the research. The recommendations included safeguards to protect individual privacy, stressing informed consent and making data anonymous where possible. Insurance companies have also been advised against accessing their clients' genetic information. The committee further suggested that a five-year moratorium could be imposed on the use of predictive genetic information for insurance purposes and that an authority could be appointed to monitor developments in this area. Having such steps in place would in turn prevent ethical lapses leading to leaks of people's private medical or genetic information. This was a key part of an effort to assure the public that taking part in research would not compromise their privacy.

CONTROVERSIES THAT ENHANCED AWARENESS HERE

While there have been a number of high-profile cases concerning fraudulent misconducts of scientists in the last ten years, such as that involving South Korean scientist Hwang Woo-Suk, the actual number of unethical acts that occur every year is not known.[52] Hwang was feted as a national hero when, in 2004, his research team was said to have successfully cloned a human embryo and produced stem cells from it, a technique that could one day provide cures for a range of diseases. But allegations that he used unacceptable practices to acquire eggs from human donors then faked two landmark pieces of research into cloning human stem cells left his reputation in tatters. In another case, *The Lancet*, one of the world's leading medical journal, issued in January 2010 a full retraction of discredited research linking autism and MMR, the combined vaccine against measles, mumps, and rubella, following a ruling

[51] Chen HF. Biomedical Research Data Off-limits to 3rd Parties: BAC, *The Business Times*, 15 June 2006, p. 14.
[52] Kesava S. As Research Grants Grow, Watch Out for Fraud, *The Straits Times*, 23 July 2008, pp. 1–2.

by the UK General Medical Council.[53] In Singapore, the ethical controversy over the research practices of Professor Simon Shorvon in 2002 received wide publicity.[54] Often, ethical failings are attributed to "the failure of peer review and the ever mounting pressure to publish and perform at the highest level".[55] Hence changes to the peer review process tend to follow from an occurrence of ethical misconduct. As Shobana Kesava observes,[56]

> *Here [in Singapore], Simon Shorvon stands as the only high-profile scientist whose ethics were called into question. As chief of the National Neuroscience Institute in 2002, he was found guilty of putting Parkinson's disease patients through tests without their informed consent. Since that case, hospital clusters SingHealth and the National Healthcare Group (NHG) have tightened safeguards; their respective review boards and clinical trials compliance panels keep things on the straight and narrow. SingHealth's assistant chief executive officer of research and education, Professor Soo Khee Chee, said keeping log books of study data is a must. These become pieces of evidence when investigations are launched; on the flip side, they can also vindicate the wrongly accused.*

PROTECTING PUBLIC PARTICIPATION IN RESEARCH

Public support does not extend to another critical area of Singapore's life sciences push: volunteering information and tissue, or taking part in trials. Singapore lags behind countries such as Sweden in terms of actively volunteering to give samples or participating in clinical trials; and projects here consistently have difficulty acquiring sufficient numbers. In Sweden, nine in ten people typically respond to calls to volunteer for research.[57] But in

---

[53] Triggle N. Lancet Accepts MMR Study 'False', *BBC*, 2 February 2010. Available at http://news.bbc.co.uk/2/hi/8493753.stm. See also: Author unknown. Medical Journal Retracts Autism Paper 12 Years On, *Reuters*, 2 February 2010. Available at: http://www.reuters.com/article/idUSTRE61132920100202.

[54] Chang AL. Speak Up on Science Issues Before It's Too Late, *The Straits Times*, 11 August 2009, p. A2.

[55] Gottweis H and Triendl R. South Korean Policy Failure and the Hwang Debacle, *Nature Biotechnology* **24** (2006): 141–143, p. 141.

[56] Kesava S. As Research Grants Grow, Watch Out for Fraud, *The Straits Times*, 23 July 2008, pp. 1–2.

[57] Chang AL and Loo D. Proposed Research Privacy Laws Hailed, *The Straits Times*, 19 June 2006, p. H4.

Singapore, the participation rate has, over the past few years, been clocked at 50%. The lack of volunteers has held back critical research and, in some cases, put key projects at risk.[58] Potential research participants were cautious, even fearful about losing their privacy. This was one reason why people were slow to sign up for trials, even when approached. Not only were they reluctant to give tissue samples, they were even unwilling to give details of their medical history.

There are now safeguards in place to protect personal information used in research, and they show that Singapore is serious about protecting research volunteers' confidentiality. Safeguards include the fact that volunteers must be given full information both in writing and verbally, and all clinical drug trials must be approved by both the hospital's ethics committee or an institutional review board and the Medical Clinical Research Committee (MCRC, which has the responsibility of issuing a clinical trial certificate that is required prior to the commencement of any trial). A drug trial that is known to present a serious risk of harm or is inferior to an existing treatment will not be approved by the MCRC.[59] These institutional safeguards, together with the laws in place, are intended to ensure that clinical trials conducted in Singapore are properly controlled and the well-being of trial subjects is seen to as far as possible. With the expansion in research, many see the BAC's recent introduction of a framework to oversee ethical issues in research projects as timely. Although all hospitals here already have their own ethics panels, the framework sets common standards and guidelines for all researchers.

A hospital's ethics committee has to monitor all trials carried out by its doctors. It vets the application for a trial certificate before it is sent to the MCRC. The hospital's ethics committee will continue to review the clinical trials. When a non-drug clinical trial is carried out, it will ensure that the proposed protocol addresses ethical concerns and meets regulatory requirements for such trials.[60] And the sponsoring drug company will ensure that the doctors involved know how to carry out the trial properly. This includes flying the principal investigator to international meetings where, together with doctors from other countries involved in similar trials, he will be given a complete

[58] *Ibid.*
[59] Chen HF. Review Board a Must for Institutions Conducting Human Biomed Research, *The Business Times*, 24 November 2004, p. 1.
[60] Woo KT. Conducting Clinical Trials in Singapore, *Singapore Medical Journal* **40**, 4 (1999): 310–313.

rundown of the drug, procedures, and techniques. Doctors will be provided with any equipment that is needed for the trials. Quite often, the sponsoring company will also pay for a nurse to help out.

With the safeguards in place, Singapore experienced a surge in human biomedical research in 2002, driven by the development of science here and recognition by international pharmaceutical companies that the island republic is a viable location for clinical trial outsourcing. According to the Health Sciences Authority, which regulates clinical drug trials, the number of approved applications went up from 99 in 1998 to 195 in 2002.[61] The only time this number dipped to 160 was in 2003 because of severe acute respiratory syndrome (SARS), which made potential volunteers reluctant to leave their homes. Researchers, too, held back submissions for drug trials that year. At the NHG, the number of non-drug human biomedical research studies reviewed has increased from an average of between 190 and 210 studies a year from 2001 to 2003, to about 246 since the start of the financial year in April 2004. At the other cluster, the SingHealth Group, the number reviewed by its institutional review boards rose from 326 in 2001 to 546 in 2002. In 2003, the figure slipped to 470 because of SARS, but was expected to reach 708 in 2009.[62]

More drug companies were also conducting trials in Singapore. As Daryl Loo writes,[63]

> *...Pfizer has started a series of trials here to test and fine-tune new treatments for the elderly, the world's fastest growing market. The United States-based firm also plans to double the size of its six-year-old clinical drug trial centre at the Singapore General Hospital (SGH), its sole trial centre in Asia. Pfizer joins a growing number of pharmaceutical companies that increasingly view Singapore as having the competencies and standards to conduct such drug trials, especially in the early, more challenging phases. ...Another pharmaceutical with a drug trial base here is US-based Eli Lilly, which has a centre at the National University of Singapore.*

[61] Chen HF. Human Biomed Research on a Roll Here, *The Business Times*, 3 December 2004, p. 2.
[62] *Ibid.*
[63] Loo D. More Drug Companies Conducting Trials in S'pore, *The Straits Times*, 9 September 2006, p. 2.

> *Other pharmaceuticals firms such as Schering-Plough and Covance manage regional trials from Singapore... One of the largest clinical trials ever conducted in Singapore was for a vaccine for the rotavirus, which causes severe diarrhoea and vomiting in infants, and is a main culprit of childhood death. The four-year Phase IV trial conducted by drug firm GlaxoSmithKline concluded last year and involved more than 9,000 infant subjects. It resulted in the vaccine Rotarix being approved for use in Singapore.*

Despite the surge in clinical trials here, more needs to be done to educate people about the benefits of active participation in research, and to explain to potential volunteers what each project is about, the possible risks, and what is being done to protect them.

CONCLUSION

The Singaporean public lacks knowledge and tends to trust the Government to make the right decisions in the complex field of bioethics. Its stance is reactive rather than proactive: public feedback is heard mainly when controversies happen. To build on the relationship of public trust and take it to the next level of active participation, there must be more collaborative relationships between researchers, the BAC, and the public.

But getting people here truly interested in science may be as difficult as changing Singapore's social fabric. Only when they begin to feel that science is as exciting as shopping or football, and when debating scientific issues becomes as relevant to them as tracking housing prices, will the changes in attitude happen. Then, society will be truly empowered to use science for what people want in life.

4

Confucian Trust and the Biomedical Regulatory Framework in Singapore

Anh Tuan Nuyen

Outsiders may be forgiven for thinking that Singapore does not have a very robust legal framework regulating biomedical research. It is true that compared with Western Europe and North America, there is little legislation governing research practices in the biomedical field. (To date the two relevant pieces of legislation that relate directly to biomedical research are the *Medical (Therapy, Education and Research) Act* and the *Human Cloning and Other Prohibited Practices Act*.)[1] This is somewhat surprising given the fact that the Government has poured considerable resources into biomedical science, which have contributed to a burgeoning growth in the teaching and research in this area, as well as its commercialisation. To be sure, the Government has set up a Bioethics Advisory Committee (BAC), the remit of which is to "address the ethical, legal and social issues arising from biomedical sciences research in Singapore" and, after actively gathering information and views from the international and local communities and careful deliberation, to "recommend policies to the Steering Committee on Life Sciences". Since its inception in December 2000, the BAC has issued six reports with recommendations covering a wide range of biomedical research practices.[2] However, very few of those recommendations have been taken up in legislation.

It is fair to say that there is in Singapore a legal 'light touch'. Observers may well ask whether the 'light touch' is the result of deliberate policy, or the low

[1]The Medical (Therapy, Education and Research) Act 2010 (as amended), Chapter 175, and the Human Cloning and Other Prohibited Practices Act 2004, Chapter 131B.
[2]The reports of the Bioethics Advisory Committee are available at: www.bioethics-singapore.org.

legislative priority being assigned to this area of the law. If the former, can the policy be justified? If the latter, should a higher priority be assigned? These questions will not be directly addressed in this paper. Instead, I will argue that given a certain cultural background, a case can be made for the 'light touch'. I will argue that in a cultural environment where there is an emphasis on individualism and individual rights, there is a tendency to resort to law as a protection of rights. By contrast, in a cultural environment where there is an emphasis on trust, there is less need to resort to law to regulate social conduct. I will show that since trust, or trustworthiness, is a key Confucian virtue, a country with a cultural environment that is broadly Confucian can justify the legal 'light touch' in certain areas. It does not follow that non-Confucian societies, such as 'the West', are less inclined to do so or do not have the conceptual resources to do so. However, in the case of the West, the reliance on trust can only come about if the tension between trust and individual autonomy is satisfactorily resolved. Since there is no such tension in Confucian thinking, as I will show, we can at least understand why a society that aspires to be Confucian may opt for the legal 'light touch' as a deliberate policy in certain areas of social practices.

INDIVIDUAL AUTONOMY, RIGHTS, AND THE LAW

According to Onora O'Neill, autonomy and trust, as they are typically conceived, are incompatible with each other:[3]

> *Trust flourishes between those who are linked to one another; individual autonomy flourishes where everyone has 'space' to do their own things. Trust belongs with relationships and (mutual) obligations; individual autonomy with rights and adversarial claims.*

It is no wonder, she goes on to observe, that "amid widespread and energetic efforts to respect persons and their autonomy and to improve regulatory structures, public trust in medicine, science and biotechnology has seemingly faltered".[4] O'Neill, of course, is not suggesting that the rise in efforts to respect persons and their autonomy is the cause of faltering public trust. Indeed,

[3] O'Neill O. *Autonomy and Trust in Bioethics.* Cambridge: Cambridge University Press, 2002, p. 25.
[4] *Ibid.* p. 3.

she suggests that there are "some conceptions of autonomy and trust" under which they are "compatible, and even mutually supporting".[5] However, even if O'Neill succeeds in showing this, the fact remains that the conceptual focus of autonomy is still the individual and that of trust the community, and that autonomy and trust have historically played different roles in the way we actually relate to each other and in our reasoning concerning how we ought to do so.

Historically, the West has moved steadily towards confirming the supremacy of the individual. Personhood is now understood as something manifested in the individual, and respect for persons is just respect for individual persons. The defining feature of an individual person is his or her freedom, understood as the autonomy to act. This is not only affirmed in the Western philosophical tradition, by thinkers ranging from Descartes to Kant to contemporary philosophers, it is something that countless political struggles have established. It follows that to respect an individual person is to respect his or her autonomy, and since autonomy is exhibited in actions, it is to respect a person's actions. This entails giving a person the greatest scope possible to act and not interfering with a person's freedom to act. However, since a person's actions are more likely than not to impact on others, it is necessary to ensure that we all can have sufficient "space to do our own things", to borrow O'Neill's words. One way of doing so is to accord to each individual a set of rights. Defenders of rights argue that a person's rights define the space in which he or she acts, and to violate any right is to invade that space, thus failing to show respect for the person. Rights, in turn, are typically protected by law.

In the past few decades, the world in general, and the West in particular, has shifted away from the idea of acting for the greater good to taking rights seriously.[6] The West is now accustomed to the idea that a law-governed society is necessary to protect individual rights and to safeguard freedom and autonomy, and has thus actively promoted this idea through international organisations. In 1976, the *International Bill of Human Rights*,[7] based on the *Universal Declaration of Human Rights* issued soon after the inception of the

---

[5] *Ibid.* p. 17.

[6] See, for instance, Dworkin R. *Taking Rights Seriously.* Cambridge, MA: Harvard University Press, 1977.

[7] United Nations, *International Bill of Human Rights*, GA A/RES/3/217, 10 December 1948.

United Nations,[8] was adopted and became international law. Incorporated in this Bill are the two "Universal Covenants" on "Civil and Political Rights" and "Economic, Social and Cultural Rights". In 1990, the *Convention on the Rights of the Child* (based on the 1928 League of Nations' *Declaration of the Rights of the Child*) was adopted.[9] In 2007, the United Nations declared the *Rights of Indigenous Peoples.*[10] In the realm of world politics, how well a country is perceived by the international community is measured by the extent to which human rights are respected. The adoption of a "Bill of Rights" is the mark of international respectability for any nation. Socially, more and more rights are identified and demanded, ranging from women's rights to animal rights. In the biomedical arena, debates are dominated by claims of rights, such as the right to life and the right to die and the right to reproductive choices and so on. In the specific case of biomedical research, it is now standard to recognise and uphold the rights of the patient and the research participant, such as the right to withdraw from a research project without having to give any reasons, the right to privacy and confidentiality, and above all the right to give or withhold consent, which must be both voluntary and informed.

Proponents of all these rights typically insist that they be protected by law, rather than just declared or otherwise acknowledged. They typically regard the legal 'light touch' as a sign of a lack of seriousness about the protection of rights and, indirectly, about respect for persons. Law is preferred where people and organisations are not trusted to do the 'right thing'. A robust legal framework in the case of biomedical research is preferred when, recalling O'Neill's observation cited earlier, "public trust in medicine, science and biotechnology has seemingly faltered".[11]

The rights-based approach to biomedical research is intuitively appealing. As mentioned above, defenders of rights believe that rights, particularly human rights, are what we need to protect and enhance personal autonomy,

---

[8]United Nations, *Universal Declaration of Human Rights,* GA res. 217A (III), UN Doc A/810, p. 71 (1948).

[9]United Nations, *Convention on the Rights of the Child,* GA A/RES/44/25, 12 December 1989; and *Geneva Declaration of the Rights of the Child,* GA res. 1386 (XIV), 14 UN GAOR Supp. (No. 16), p. 19, UN Doc A/4354.

[10]A copy of the United Nations *Declaration on the Rights of Indigenous Peoples,* GA A/RES/61/295, 2 October 2007.

[11]O'Neill O. *Autonomy and Trust in Bioethics.* Cambridge: Cambridge University Press, 2002, p. 3.

or to advance the ideal of a human person as a free agent. However, the rights-based approach is unsatisfactory in many different respects. The link to autonomy, or freedom, loses its appeal when it is realised that freedom needs to be circumscribed by responsibility. More importantly, much is lost when we insist on our rights in our dealings with others. To give something to someone because he or she has the right to it is to do so either without feelings, or with grudge. To take something from someone on the basis of rights is to do so without gratitude and without appreciation. According to Peter Williams, gratitude and praise are moral sentiments which are inappropriate when rights are asserted, a reason why for him "a moral universe in which our only claims on each other were claims of right would be morally impoverished. It would be devoid of thanksgiving or praise and without celebration".[12] Think of the difference between parents who care for their children out of love and those who do in observation of the United Nations *Declaration of the Rights of the Child*!

All this can be explained by the fact that rights protect a person by setting up barriers, typically legal ones, around this person, also separating him or her from others. To make a rights claim is to assume a combative position, to put oneself in an adversarial position *vis-à-vis* others. A rights claim is always a claim *against* someone. The language of rights is replete with terms that suggest conflicts and violence, such as 'violation', 'transgression', 'infringement', and so on. To be sure, we should be thankful that there are such things as the United Nations *Declaration of Human Rights, Rights of the Child,* and other rights-based declarations. However, such declarations are born out of degradation, oppression, injustice, and inhumanity. Thus, rights claims are associated with the dark side of humanity and such things as the United Nations *Declaration on Human Rights* exist as a matter of relief. To engage with others 'armed' with rights is to do so with suspicion and with a lack of trust. As such, rights claims are not conducive to co-operative relationships and do not promote solidarity.

This is a considerable disadvantage in the case of biomedical research. The complex and technical nature of biomedical research requires trust and co-operation; its noble goal of serving humanity requires a positive spirit, not something that reminds us of conflicts and violence; its accomplishments are

[12]Williams P. Rights and the Alleged Rights of the Innocents to Be Killed, *Ethics* **87** (1977): 384–394, p. 394.

a matter of thanksgiving, praise, and celebration. All this is not to say that the interests of patients and research participants should not be protected, in the case of biomedical research or generally, nor is it to say that freedom, autonomy, and respect for persons do not matter. It is to say only that there is a need to explore a different approach to protecting the interests of all concerned, to safeguard freedom and autonomy and to show respect for persons.

PERSONS AS RELATIONAL SELVES AND TRUST

Arguably, the emphasis on individual rights as a promoter of personal autonomy is the outcome of the metaphysical conception of a person as an autonomous individual, a self, independent of all other co-existing individuals, other selves. As is well known, Descartes encourages us to encounter the 'true self' by stripping away all that is contingent about one's self. This true self, being a self-sufficient, autonomous individual who understands himself or herself as separate from and independent of others, chooses to form relationships with others, relationships that are purely contingent, and makes choices as a rational, autonomous, and independent self.

This conception of a person stands in contrast with the Confucian conception. To be sure, commentators do not agree on what exactly the latter is, and indeed some interpretations of it seem implausible. Arguably, the most plausible characterisation of the Confucian person is something like Chenyang Li's. Li contends that in "the Confucian view, [the] self is not an independent agent who happens to be in certain social relationships".[13] Rather, the self "is constituted of, and situated in social relationships".[14] In this view, the Confucian self's identity is not arrived at or defined independently of the society in which the self finds itself. Rather, the self defines itself in terms of the social roles it occupies in the society. The society, in turn, is constituted by distinct individual selves acting in different roles. Roger Ames and David Hall have helpfully employed the focal-field metaphor to characterise the relationship between the individual self and the social network of relationships in which it finds itself. For Hall and Ames, the Confucian self is "a focal [*sic*] in that it

---

[13] Li C. *The Tao Encounters the West.* Albany: SUNY Press, 1999, p. 94.
[14] *Ibid.*

both constitutes and is constituted by the field in which it resides", the "field of social activity and relations".[15]

The Confucian conception of the self has a distinct implication for ethical conduct. For Confucians, social relationships are characterised by social positions, or roles, and social positions are defined in terms of obligations. To each role is attached a set of obligations, and to be in a role is to be under a set of obligations. Which obligations go with which role is determined by more or less explicit social expectations. The key social roles are encoded in the rites, *li*. To be in a social relationship, then, is to stand under certain obligations. What one ought to do and how one ought to behave in a certain relationship are all set out in *li*, or in social expectations. *Li* describe both the factual and the ethical. Thus, in Confucian ethics, moral rules concerning duties and obligations, and the moral virtues, are all derived from the roles that define an individual as a person, or an agent. To become a superior person is to learn about and to live up to the moral requirements attached to the social roles in terms of which one is defined. It is to discharge the duties and obligations and to cultivate the virtues entailed by the social roles that one occupies, which for the Confucians constitute the true self. In Confucianism, then, the self is conceived primarily as a socially responsible agent, in contrast to the traditional moral theories in which the self is conceived as an independent person in the way mentioned above, whose autonomy is typically thought to be guaranteed by rights.

If I am right in my reading of Confucianism above, it is clear that Confucian ethics is a role-based ethics in which moral rules and virtues are derived from the social roles that an agent sees himself or herself occupying, which is consequent upon seeing his or her self in terms of a network of social relationships. While the Confucian role-based ethics stands in sharp contrast with traditional Western ethical theories, it should be noted that there are many contemporary philosophers in the West, whose views converge on a kind of ethical theory that bears many striking similarities with the Confucian account. For instance, in *Sources of the Self*, Charles Taylor argues that "I am a self only in relation to certain interlocutors", that a "self exists only within . . . webs of interlocution".[16] All this is clearly reminiscent of the

[15] Hall D and Ames R. *Thinking from the Han: Self, Truth and Transcendence in Chinese and Western Cultures*. Albany: SUNY Press, 1998, p. 43.
[16] Taylor C. *Sources of the Self*. Cambridge, MA: Harvard University Press, 1989, p. 36.

account of the Confucian self above. To be sure, the strands that make up the "webs of interlocution" identified by Taylor are not Confucian and a Confucian is likely to be completely lost in Taylor's 'geography' of 'defining relations'. Nevertheless, his moral philosophy, like Confucian ethics, is not based on deontological or teleological rules, nor on any understanding of virtues arrived at independently of the conception of a person, but rather on the roles that go with the "social statuses and functions" that define a person.

Taylor is not alone in his approach to moral philosophy. We can find the same approach in Dorothy Emmet before him. In *Rules, Roles and Relations,* Emmet follows the same line that has been delineated in Confucianism above and argues that the self is not an autonomous individual independent of others, but rather a person who stands in a "nexus of relationships".[17] Morality and moral judgements are based on "structured commitments and expectations within a network of relationships".[18] As in Confucian ethics, in Emmet's account, rules in general, and moral rules in particular, are based on social roles, which in turn are defined by social relations. Emmet defines a role as "a capacity in which someone acts in relation to others",[19] and claims that "a role relation in a social situation has some notion of conduct as appropriate or inappropriate built into its description".[20] In language that is clearly reminiscent of Confucianism, Emmet argues that in a statement such as "You ought to help her because, after all, she is your mother", the "obligation to help is said to follow from the fact of parenthood". She goes on: "But the fact is not a mere fact; it is a fact of social relationship. And a fact of social relationship is one about people occupying roles *vis-à-vis* each other."[21]

Emmet and Taylor are not two isolated thinkers coincidentally converging on a view of ethics that bears comparison with Confucian ethics as role-based ethics. It can also be found in the works of many other thinkers. Marion Smiley,[22] for instance, contrasts her view on moral responsibility with what she calls the "prevailing view", or the "modern concept of moral responsibility". In her view, responsibilities are directly or indirectly role-based. Thus,

[17] Emmet D. *Rules, Roles and Relations.* London: Macmillan, 1966, p. 139.
[18] *Ibid.* p. 201.
[19] *Ibid.* p. 13.
[20] *Ibid.* p. 15.
[21] *Ibid.* p. 40.
[22] Smiley M. *Moral Responsibility and the Boundaries of Community: Power and Accountability from a Pragmatic Point of View.* Chicago: University of Chicago Press, 1992.

we impose them, or attribute them, to others "on the basis of the variety of normative expectations that are themselves grounded not only in our configuration of social roles, but in the interests, power relations, and structures of community that support such roles in practice".[23] In a passage reminiscent of Taylor and Emmet, Smiley argues that responsibility and blameworthiness have to be accounted for in terms of 'social expectations', which evolve "within a complex web of practical considerations, traditional beliefs, and configurations of social roles".[24] To be sure, Smiley's concern is specifically with the duty to relieve the suffering of others. However, there is no reason why the ethical foundation she has constructed to account for it cannot be used to ground other duties in "our practical life".

Finally, Larry May in *The Socially Responsive Self*[25] begins his account of moral responsibility in the same way Emmet does, namely with "the self as agent" in Emmet's words. May's self turns out to be remarkably like the Confucian self I described above, being "largely a product of social factors".[26] Like Emmet, May, in an earlier work, contends that "Social roles generally increase the domain of responsibility for those who assume or who agree to be cast into these roles."[27] He goes on: "Role responsibilities are responsibilities one has by virtue of having agreed to take on a certain set of tasks in society, or perhaps by virtue of having agreed to be thrust into the position of assuming various tasks."[28] As if having Confucianism in mind, May claims that communal harmony "is best sustained when all of the members of a community are striving to advance the common good" in playing their roles.[29]

Given the metaphysical conception of the self as a person standing in a network of social relationships, found in Confucian thinking as well as in the thinking of many Western philosophers, trust takes on a significant and indispensable role in ethical reasoning as well as in social practices. O'Neill, as we saw earlier, has rightly observed that trust "flourishes between those who are linked to one another . . . [and] . . . belongs with relationships and

[23] *Ibid.* p. 185.
[24] *Ibid.* p. 190.
[25] May L. *The Socially Responsive Self: Social Theory and Professional Ethics.* Chicago: University of Chicago Press, 1996.
[26] *Ibid.* p. 3.
[27] May L. *Sharing Responsibility.* Chicago: University of Chicago Press, 1992, p. 163.
[28] *Ibid.* p. 164.
[29] *Ibid.* p. 171.

(mutual) obligations". Indeed, we can say that the sense of "belonging" here is a conceptually strong one, at least for many key social relationships, which presupposes trust. O'Neill herself mentions "the doctor-patient relationship" as "a paradigm case of a relationship of trust".[30] However, if we are to maintain the contrast between rights and trust, relationships such as this become problematic. Indeed, in many legal systems, relationships of this kind are regulated by law, which imposes *fiduciary duties* on certain parties. Doctors, lawyers, bankers, etc. have fiduciary duties, or duties of trust, to act in the best interest of patients and clients and failure to do so constitutes a legal breach of trust.

In these relationships, legal rights and trust go together. If we are to argue for a shift from the rights-based approach to one based on trust, then we should set to one side cases involving fiduciary duties and focus on relationships such as friendship, filial love, conjugality, companionship, comradeship, fellowship, mateship, and so on. It is in these relationships that claims of rights are at best unhelpful and at worst destructive. By contrast, it is trust that is the foundation on which these relationships are built. However, it is a different kind of trust from the trust that is integral to fiduciary duties, the latter being tied to rights and guaranteed by law. In the relationships in question, trust is grounded not in rights or law but in trustworthiness, or in the character of the trusted person, or more deeply in his or her *moral* character, not in his or her professional capacity.

For Confucians, a society consists of relationships built on the kind of trust that is grounded in trustworthiness. A self is a person standing in a network (or nexus, to use Emmet's term) of such relationships. This is why trustworthiness, *xin*, is a key Confucian virtue. Indeed, together with humanity, *ren*, propriety, *li*, righteousness, *yi*, and wisdom, *zhi*, it makes up the five Confucian constant, or cardinal, virtues. Confucius teaches it as one of the "four subjects",[31] regarding it as one of the three ways of exalting morality[32] and as the virtue that yokes all other virtues together like the yoke that holds together the

[30]O'Neill O. *Autonomy and Trust in Bioethics*. Cambridge: Cambridge University Press, 2002, p. 17.
[31]*Analects* 7.25. Readers wishing to consult Confucius' *Analects*, or other works of Chinese philosophy, may find a useful resource to be Chan WT. (comp. and trans.) *A Source Book in Chinese Philosophy*. Princeton, NJ: Princeton University Press, 1963.
[32]*Analects* 12.10.

component parts of a cart.[33] Thus, the moral person must aim at cultivating trustworthiness. Indeed, for Wei-ming Tu, the ideal Confucian society is a society governed by trust, or a fiduciary community of trustworthy people.[34] The Confucian emphasis on trust is to be expected since Confucians conceive the self as a person standing in a network of relationships, and since the foundation of social relationships is trust. Both Confucius and Mencius take it for granted that the trustworthy person will be trusted.[35] Since many social relationships are mutual, the trust in question is a mutual trust. Also, the more those involved in these relationships are trustworthy, the more trust there is. In turn, the more trust there is, the stronger the social relationships that are built on it and the more cohesive the community that consists in such relationships.

In a robust relationship, one party trusts the other to discharge the obligations and responsibilities entailed by such a relationship. True friends trust each other to 'do the right thing' as friends, standing by each other, keeping the other's confidence, and so on. The Confucian idea of cultivating the self consists in learning what the social relationships entail and living up to what is entailed in terms of obligations and responsibilities. This is also known as the rectification of names, the process of learning about and living the social roles and relationships that define oneself as a person. The person who has cultivated himself or herself well is a trustworthy person, who can be trusted with all the responsibilities that a role or relationship entails. Thus, Confucius says that if a man "is trustworthy then his fellow men will entrust him with responsibility".[36] It follows that where there is trust, there is less need for rules and regulations to ensure that people act responsibly. Rules and regulations may certainly force people to act responsibly but Confucius believes that virtues in general and trustworthiness in particular would be far better and far more effective as the force that motivates people to act appropriately. In particular, where there is trustworthiness, there is trust — mutual trust in the case of many key social relationships — and where there is trust and mutual trust, the trusting people are less inclined to feel that their interests have to be protected by law as a matter of rights. Famously, Confucius says: "Lead the people with governmental measures and regulate them by law and punishments and they will avoid wrongdoing but have no sense of honour or

[33] *Analects* 2.22.
[34] Tu WM. *Centrality and Commonality: An Essay on Confucian Religiousness.* Albany: SUNY Press, 1989.
[35] *Analects* 17.6 and *Mencius* 7B12 (see also footnote 31).
[36] *Analects* 17.6 — English translation adapted from various sources.

shame. Lead them with virtue and regulate them by the rules of propriety (*li*) and they will have a sense of shame and moreover set themselves right."[37] To be sure, the virtue mentioned in this passage is *li* but, as mentioned earlier, all virtues, including *li*, are 'yoked' to a person by the virtue of trustworthiness, *xin*, according to Confucius. It is true also that this passage refers to the behaviour of the ruler but it clearly applies to each person individually and more importantly says that 'governmental measures and law' are inferior means of regulating behaviour.

Thinkers such as Taylor and Emmet, even though they have not explicitly emphasised the role of trust, would no doubt endorse the view that where there is trust grounded in trustworthiness, there is less need to insist on rights and to resort to law as a protection of rights. If parents are trustworthy as parents, we can trust them to carry out their parental duties appropriately. In a society where parents are parents, as Confucians would say,[38] there is less need for laws to ensure that the "rights of the child" are protected. This is not to say that we can do away with rights, or with laws altogether. After all, even Wei-ming Tu concedes that the fiduciary community is a Confucian *ideal*, not necessarily a reality in any Confucian society.[39] Rights may still have to be defended by laws and regulations on the way to the ideal. However, in the Confucian vision, as in the moral vision of thinkers such as Taylor and Emmet, moral progress is not one towards a rigidly regulated society but rather one towards a society that consists of members who understand themselves not as isolated and independent individuals who need to be protected from other individuals, but rather as selves who stand in networks of social relationships, which are built on the kind of trust that is grounded in trustworthiness. For any actual society, this vision may be realised in isolated areas. Arguably, in the area of scientific research in general and biomedical research in particular, where researchers are highly educated and well trained and where the goals of the endeavours, namely the advancement of knowledge, the saving of lives, and the curing of illnesses, are lofty and noble, we can expect to be closer to the envisioned ideal than elsewhere. If this is right, then the legal 'light touch' is a defensible deliberate policy.

[37] *Analects* 2.3.

[38] This is a reference to a passage in *Analects* 12.11, where Confucius says: "Let the ruler be ruler; the minister, minister; the father, father; and the son, son."

[39] Tu WM. *Centrality and Commonality: An Essay on Confucian Religiousness.* Albany: SUNY Press, 1989.

What about autonomy? Isn't the enforcement of rights by law a necessary requirement to ensure a person's autonomy and the respect that one deserves as an autonomous person? In reply, we can remind ourselves first of all that respect is earned rather than conferred by law. A trustworthy person, the Confucian *xin ren*, has justly earned our respect and will most likely be respected and trusted. More importantly, it can be argued that the Confucian moral vision is in no way incompatible with autonomy. In this vision, the moral agent is utterly autonomous in the cultivation of the self and in the rectification of names, that is, in learning to live the social roles entailed by the relationships that define the self. Perhaps the trust that is part of this vision is incompatible with a narrow kind of autonomy. O'Neill has made a distinction between what she calls "individual autonomy" and "principled autonomy": the former is "autonomy as independence"[40] and the latter is "Kant's distinctive conception of autonomy",[41] which she describes as "*autonomy of reason, ... autonomy of ethics, ... autonomy of principles* and *autonomy of willing*".[42] A principled autonomous agent, in her Kantian construction, is a person who is guided in his or her actions by these elements of autonomy, whereas an individualistic autonomous agent is one who acts independently of others. O'Neill then dismisses the moral significance of individual autonomy, claiming that it "may encourage ethically questionable forms of individualism and self-expression and may heighten rather than reduce public mistrust in medicine, science and technology", and as such, it "cannot provide a sufficient and convincing starting point for bioethics, or even for medical ethics".[43]

O'Neill's conception of principled autonomy dovetails nicely with the Confucian and Emmet–Taylor conceptions of the self discussed above. Like the Confucian self, or the Emmet–Taylor self, her principled autonomous self "takes *relationships* between obligations bearers and right holders, including institutionally defined relationships, as central".[44] In her moral vision, human rights are not swept aside: she does take rights seriously but seeks "to anchor an account of human rights in an account of *human obligations* (or *human*

[40] O'Neill O. *Autonomy and Trust in Bioethics*. Cambridge: Cambridge University Press, 2002, p. 23.
[41] *Ibid.* p. 74.
[42] *Ibid.* p. 83 (emphasis in original).
[43] *Ibid.* p. 73.
[44] *Ibid.* p. 82 (emphasis in original).

duties)".[45] There is no tension between rights and trust. Indeed, the "ethical implications" of principled autonomy "underpin the importance of trust".[46] If O'Neill is right, then the Confucian approach of regulating by trust and trustworthiness does not mean we have to downgrade human rights, autonomy, and respect for persons. It simply offers a different approach to attaining these values.

THE CASE OF SINGAPORE

As pointed out above, it would appear that the authorities in Singapore have adopted a legal 'light touch' when it comes to biomedical research. If my arguments above are correct, such an approach is understandable and even defensible. There is some evidence to show that Singapore leaders do think of Singapore as a different kind of society from the West. Leaders such as Lee Kuan Yew often speak of 'Asian values' that are distinct from 'Western values'.[47] Judging from various public comments, it may be said that the 'Asian values' that guide public policies in Singapore are grounded in a conception of the self pretty much like the Confucian conception discussed above. Thus, Singapore's Ambassador-at-Large, Tommy Koh, has been quoted as saying that the individual[48]

> *... is not an isolated being, but a member of a nuclear and extended family, clan, neighborhood, community, nation and state. East Asians believe that whatever they do or say, they must keep in mind the interests of others ... the individual tries to balance his interests with those of family and society.*

Naturally, it has to be noted that this is distinctively Asian only if we contrast it with the conception of the self as an independent person who takes his or her autonomy as the capacity to act independently of others. To be sure, this kind of individualism is often attributed to the West, indeed, the "prevailing

[45] *Ibid.* p. 78 (emphasis in original).
[46] *Ibid.* p. 74.
[47] See, for instance, Zakaria F. Culture Is Destiny: A Conversation with Lee Kuan Yew, *Foreign Affairs* **73** (1994): 109–129.
[48] Koh T. The 10 Values that Undergird East Asian Strength and Success, *International Herald Tribune*, 11–12 December 1993, p. 6.

views" as O'Neill has put it.[49] However, it is at least misleading to say that it is exclusively or typically Western, or to say that the West has no conceptual resources to conceive of a person in terms of social relationships and social obligations.

If I am right, then while a commitment to 'Asian values' is not necessary for a society to prefer to rely more on trust and less on legislation in the field of biomedical science — it has been argued for by O'Neill and would certainly be suggested by Emmet, Taylor, and other authors discussed above — such commitment is certainly sufficient to explain the preference for the legal 'light touch'. The more important question is whether, as a matter of public policy, the legal 'light touch' can be justified. Critics might want to argue that the scientific scandals coming out of some Asian countries, such as Korea, indicate the need for comprehensive legislation. It may be possible to respond by saying that some Asian countries are only *geographically* Asian, not culturally so. However, this is at best disingenuous. There has to be a better response, but before one is given, it has to be said that the legal 'light touch' does not mean 'no touch'. It is not suggested here that a Confucian society can afford to do away with law or, more particularly, to adopt a 'free for all' approach concerning biomedical science. The question is only how extensive the law should be.

Two things can be said in response to the critique of the legal 'light touch'. First, the fact that there is law does not guarantee that there will be no law-breakers. (Indeed, cynics are apt to say that the converse is true: there are law-breakers precisely because there is law!) The question is whether law or trust is a more effective way of ensuring proper conduct (in the context of biomedical research, which is the focus of this paper). Confucius may have been too optimistic in claiming, in the passage cited above, that "regulate them by law and punishments and they will avoid wrongdoing".[50] In any case, he certainly believes that it is just as effective, and morally better, to let people be guided by virtues, such as trustworthiness. Second, the legal 'light touch' is not an encouragement of conduct that violates rights, such as the rights of research participants (to privacy, to informed consent, etc.). The question is whether trustworthiness or legislation is the more effective tool to

[49]O'Neill O. *Autonomy and Trust in Bioethics.* Cambridge: Cambridge University Press, 2002, p. 23.
[50]*Analects* 2.3.

ensure that the interests of all parties are protected, whether it is better that researchers respect persons and their rights as a matter of obligation or as required by law. In any case, it all boils down to trust in the end, whether it is trust in the trustworthiness of scientists and administrators, or trust in the law, which is of course short for trust in the trustworthiness of those who administer the law.

Naturally, it remains a question how to inculcate trust and cultivate trustworthiness. The Confucians have a clear answer: the cultivation of the self. In particular, it is the cultivation of the self as a person standing in a network of social relationships, whose identity is constituted by how one is related to others in the society and thus by the obligations, duties, and responsibilities that are definitive of such social relations. In practical terms, this could well entail a certain kind of scientific curriculum, one that emphasises science as a social enterprise, not an individual pursuit, and the role of the scientist as entailing certain obligations to the society, not just the narrow obligations to the scientific community. It would appear that O'Neill has a somewhat different suggestion. Interestingly, she observes that "more formalized procedures may deepen . . . distrust . . . ",[51] and that "some of the *most* intensely regulated areas of medicine, science and biotechnology enjoy least public trust".[52] Instead, she believes that trust and trustworthiness can be cultivated in a "more open public culture" where "information is available to the public".[53] In the case of biomedical science, she suggests greater "Public consultations and public hearings",[54] "increased public participation",[55] and "extra public discussion, consent and endorsement".[56] Perhaps the main difference between O'Neill's approach and the Confucian one lies in the fact that for O'Neill, "trustworthiness and trust pull apart"[57] but "increased public participation" will bring about both, "will not only make officials and experts more trustworthy, but will eventually revive public trust",[58] whereas for Confucians, trustworthiness and trust come together and are generated from the same process of cultivating the self. It is beyond the scope of this paper to judge the

[51] O'Neill O. *Autonomy and Trust in Bioethics.* Cambridge: Cambridge University Press, 2002, p. 130.
[52] *Ibid.* p. 138 (emphasis in original).
[53] *Ibid.* p. 134.
[54] *Ibid.* p. 137.
[55] *Ibid.* p. 138.
[56] *Ibid.* p. 170.
[57] *Ibid.* p. 165.
[58] *Ibid.* p. 138.

comparative merits of O'Neill's and the Confucian views on the relationship between trustworthiness and trust. Suffice it to say that in Singapore, the BAC certainly endorses 'increased public participation' and 'public discussion', and conducts numerous "public consultations and public hearings" in the process of formalising its recommendations. Thus, it may be said that Singapore relies on a two-prong approach to ensuring trustworthiness and trust: the O'Neill prong of 'increased public participation' and the Confucian prong of cultivating 'Asian values'.

5

The Clinician-Researcher: A Servant of Two Masters?

Alastair V Campbell, Jacqueline Chin and Teck Chuan Voo

INTRODUCTION

Since the first formulation of the Helsinki Declaration it has been recognised that a tension exists between the obligations of physicians as researchers and their role as guardians of the patient's health and welfare. The horrific revelations in the Nuremberg Trials of the atrocities committed by German doctors (atrocities matched, as it emerged later, by Japanese doctors) had led to the formulation of the Nuremberg Code to ensure that the consent and protection from harm of human experimental subjects could never be disregarded. However, a period of complacency followed among medical researchers in both the USA and the UK, who supposed that such inhumanity could never be ascribed to their rapidly burgeoning research endeavours. This complacency was shattered in the mid-1960s by the revelations of Henry Knowles Beecher[1] in the US and Maurice Henry Pappworth in Britain, demonstrating that a whole range of studies showed — at best — a casual disregard for patient consent and no real estimate of risks.[2] (The illusion was finally dispelled in the US when details of the Tuskegee Syphilis Study emerged, a federally funded study that allowed sufferers to die of a treatable disease in the interests of studying its natural history.) The

[1] Beecher HK. Ethics and Clinical Research, *New England Journal of Medicine* **274**, 24 (1966): 1354–1360.

[2] Pappworth MH. *Human Guinea Pigs: Experimentation on Man*. London: Routledge & Kegan Paul, 1967.

World Medical Association, meeting in Helsinki in 1964, adopted the first of many versions of a set of guidelines designed to ensure that all doctors were left in no doubt regarding their obligation to gain informed consent and to ensure that "the health of their patients was their first consideration" (this formulation taken from the 1948 *Geneva Convention Code of Medical Ethics*).[3]

Yet, even in this new era of ethical awareness, there remained uncertainties about the dual obligations of doctors. Some influential physicians fought a rearguard action against the proposed requirement for informed consent in clinical settings,[4] and the effect of this can be seen in the way in which the first version made a distinction between 'therapeutic' and 'non-therapeutic' research, with some greater leeway on consent in the former. (We shall return below to this 'therapeutic exception' and the ways in which it could be misinterpreted, indeed abused.) The distinction was eventually abandoned and later versions are much clearer on the hazards of the dependent relationship between doctor and patient when enrolment in trials is being sought. The current (2008) version of the Helsinki Declaration recommends independent consent taking:[5]

> *When seeking informed consent for participation in a research study the physician should be particularly cautious if the potential subject is in a dependent relationship with the physician or may consent under duress. In such situations the informed consent should be sought by an appropriately qualified individual who is completely independent of this relationship.*

In its 2004 *Report on Research Involving Human Subjects*, the Singapore Bioethics Advisory Committee (BAC) acknowledged the potential conflicts of interest (COI) and endorsed the above requirement for independent consent taking.

---

[3]Second General Assembly of the World Medical Association, *Declaration of Geneva*, 1948.
[4]For example, Austin Bradford Hill, writing in the *British Medical Journal* in 1963, argued that patients were not capable of understanding the complexities of medical research and would lose confidence in their doctors if they were compelled to give such explanations; Hill AB. Medical Ethics and Controlled Trials, *British Medical Journal*, 1 (1963): 1043–1049.
[5]World Medical Association, *Declaration of Helsinki — Ethical Principles for Medical Research Involving Human Subjects*, 22 October 2008 (as amended), para 26.

So is all now well, with no real worries about 'serving two masters'? We fear not! There are a number of factors influencing current medical and research practice, which may well militate against even the best of intentions of the medical profession and the research community to safeguard the freedom and safety of patients. The first is the ever-increasing emphasis worldwide on translational clinical research. While this is clearly highly desirable in many respects, it greatly increases the pressure on clinicians to combine practice and research, rather than focusing solely on treating patients. Career paths are being formulated that tie promotion to research output as much as to clinical excellence. There is a risk that such clinician-researchers, not necessarily fully versed in the dilemmas of research ethics, will fail to see the risky ambiguities in their relationships with patients. Allied to this are the widely recognised pressures on both institutions and individual physicians to enlist their patients in pharmaceutical trials, in which company profit may be as much a factor as patient benefit. The COI implicit in using patients for research is clearly heightened by financial incentives totally removed from questions of priorities for patient care. Thus the question remains: which master is being served?

We shall explore these issues in the rest of this chapter, first, by considering two well-documented examples in which the clinician-researcher roles had unfortunate consequences for patients; we shall then discuss in more detail the questions of the 'therapeutic misconception', COI, and failure to understand or respect ethical principles, which are raised by these cases. In the final section we shall make some brief suggestions for more effective measures to ensure that the trust of patients in clinician-researchers is truly justified.

WHEN THINGS GO WRONG: TWO CASE EXAMPLES

The first example concerns research conducted by Associate Professor Herbert Green at National Women's Hospital, Auckland, New Zealand. Although the research began in 1966, full details did not come fully into the public arena until 1987, when the New Zealand government set up a Judicial Inquiry, following revelations in a popular magazine, *Metro*. The second example is of research conducted in Singapore by Professor Simon Shorvon, a neurologist from the UK who was appointed Director of the Singapore National Neuroscience Institute (NNI) in 2000. Both examples illustrate how medical practitioners can ignore questions of patient safety and informed consent in pursuit of research objectives.

The Green Case

In June 1966 a proposal was put to the Senior Medical Staff of Auckland's National Women's Hospital by a gynaecologist, Associate Professor Herbert Green, to have all women at the hospital under the age of 35 who had positive smears for cervical carcinoma in situ (CIS) referred to him for treatment and follow-up. Green believed that this form of cancer was not always a precursor to fully invasive cancer of the reproductive tract and that the current accepted treatment in Auckland and worldwide was unnecessarily invasive. (The standard treatment was a procedure called cone biopsy, whereby all of the affected tissue was removed for histological examination and the women were then tested regularly to ensure that no cancerous cells remained.) Instead of this standard treatment Green proposed to do the lesser procedure of punch biopsy (a purely diagnostic procedure which would not remove all the affected tissue) and then to follow up on the women with repeated tests and clinical examinations, in order to prove his hypothesis that CIS was usually benign. The minutes of the meeting stated that "in the interests of continuity of supervision and patient confidence" all such patients should be referred to Green "whose conscience is clear and who could therefore accept complete responsibility for whatever happens".[6] (Significantly, none of the private patients of Green's colleagues were referred to him for this non-standard procedure.)

We do not propose to go into all the findings of the Judicial Inquiry which was set up in 1987 and the report published in 1988. (For a full discussion of the judge's findings and of the subsequent effects on research ethics regulation in New Zealand and worldwide, the reader is referred to *The Cartwright Papers*, edited by Joanna Manning of the Faculty of Law, University of Auckland.[7]) The Inquiry revealed that many women were inadequately treated as a result of this trial as well as being subjected unnecessarily to repeated tests and clinic visits over many years, that no attempt was made to monitor the trial adequately, and that the concerns of colleagues for the safety of the women involved were either suppressed or ignored. But, in addition, the Inquiry found that there was a total lack of information to the patients about

[6] Cartwright SR. *The Report of the Committee of Inquiry into Allegations Concerning the Treatment of Cervical Cancer at National Women's Hospital and into Other Related Matters*. Auckland: Government Printing Office, 1988, p. 21.
[7] Manning J. *The Cartwright Papers: Essays on the Cervical Cancer Inquiry of 1987–88*, Wellington: Bridget Williams Books, 2009.

the true nature of Green's unorthodox approach to CIS and no informed consent to participation in his trial. In her Report the judge summarised the lack of consent in this situation as follows:

> *The fact that the women did not know they were in a trial, were not informed that their treatment was not conventional and received little detail of the nature of their condition were grave omissions. The responsibility for these omissions extends to all those who having approved the trial, knew or ought to have known of its mounting problems and design faults and allowed it to continue.*[8]

In light of this judicial finding, it is truly surprising that, 20 years after the publication of the Report, historian Linda Bryder can portray Green as a well-intentioned practitioner doing what he saw as best for his patients and following the accepted practice of his time. In an extraordinary re-interpretation of the original version of the Declaration of Helsinki, she states: "Green did not breach the 1964 Declaration of Helsinki, which contained a strong exemption for patient consent in therapeutic research."[9] Here is what the Declaration of Helsinki states: "If at all possible, consistent with patient psychology, the doctor should obtain the patient's freely given consent after the patient has been given a full explanation."[10]

A 'strong exemption'? Any normal reading of this clause would see it as *requiring* full information and consent except in unusual circumstances (for example, when the patient was uncontactable, unconscious, or unable to cope with the information). But if what Green was doing was quite reasonable and justifiable, why could he not tell his patients what his theory was and what exactly he was doing, instead of leaving them with the impression that he was following standard practice and that they need not worry about the cancer progressing? Telling them would have been distressing ('inconsistent with patient psychology') *only if* he were subjecting them to needless risk — which indeed he was! The fact that research is risky is not a reason provided by

[8] Cartwright SR. *The Report of the Committee of Inquiry into Allegations Concerning the Treatment of Cervical Cancer at National Women's Hospital and into Other Related Matters.* Auckland: Government Printing Office, 1988, p. 69.
[9] Bryder L. *A History of the 'Unfortunate Experiment' at National Women's Hospital.* Auckland: Auckland University Press, 2009, p. 69.
[10] World Medical Association, *Declaration of Helsinki: Recommendations Guiding Doctors in Clinical Research*, 1 June 1967, Section II(1).

the Declaration of Helsinki for keeping patients in the dark — on the contrary the next section of the Declaration requires clinician-researchers to pursue research only if it is justified by its therapeutic value for the patient. Thus the Helsinki Declaration does not by any stretch of the imagination excuse Green for failing to tell his patients that in fact he was not treating them at all, but merely following them up with repeated tests, in order to prove his hypothesis that CIS was a benign condition.

Thus the so-called 'therapeutic privilege' (which normally refers to concealing distressing diagnostic information from patients) in no way justifies failure to tell patients that they are enrolled in a research project nor does it sanction a decision to conceal from them the risks that this might entail. Yet even now, in the 21st century, writers like Bryder, supported in her views by some members of the medical profession, seem to believe that the clinician-researcher rises above the standard requirements of research ethics.

The Shorvon Case

A very similar form of denial of the importance of ethics is evident in the more recent Shorvon case. Shorvon was appointed as Director of the NNI in 2000, having been recruited from the UK, where he already had a distinguished career as a researcher in neuroscience. In 2002 he was awarded a major grant by the Singapore government to investigate genetic factors influencing disease susceptibility and drug responsiveness in three of Singapore's ethnic populations (Chinese, Malay, and Indian), in respect of common neurological conditions, including Parkinson's disease. Following complaints from some patients and their doctors, a series of enquiries was launched, culminating in a hearing before the Disciplinary Committee of the Singapore Medical Council (SMC), which found him guilty of 30 charges of professional misconduct. Shorvon did not contest these charges, and he chose not to appear at the hearing, having requested de-registration from the SMC's Register. He returned to the UK, where he remains registered with the General Medical Council (GMC). The GMC considered whether it should also institute a case against Shorvon, but decided not to proceed. A judicial appeal by the SMC against this decision of the GMC was dismissed by an English court.

The essence of the case against Shorvon was that he failed to obtain informed consent for the research from several of the subjects, inappropriately obtained the names of potential participants (through bypassing their own doctors and going to pharmacy records), and failed to inform the relevant ethics committees of some of the crucial details of his research. Of particular significance here was a test which involved instructing patients with Parkinson's disease to omit their medication the evening before coming to the clinic ('on-off' levodopa testing) and then videotaping them to observe changes in their movements and co-ordination. It emerged in the SMC hearing that neither the patients nor the ethics committees were properly informed of these aspects of the research, and that patients were under the false impression that this was part of treatment, not a research test that had no relevance to their treatment.

Of especial interest in this case is an article subsequently published by Shorvon in *The Lancet* entitled 'The Prosecution of Research — Experience from Singapore'.[11] In this paper he claims that Western researchers are at serious hazard in going to such countries, because of their disregard of human rights (in this case, his!). This is an extraordinary claim, when we consider that the facts of the case are not in dispute — even a British neurological colleague giving evidence to the GMC to persuade them (successfully) not to investigate the case did not contest any of the facts described above. Instead, he described the breaches as "relatively minor . . . and in the context of increasingly bureaucratic ethical boards". Shorvon's *Lancet* paper repeats this claim that he was a victim of bureaucratic overreaction — but also claims that the whole procedure was solely politically motivated. This is mere innuendo, without any evidence to back it up. A full record of the SMC proceedings is available and shows a clear process very similar to those of disciplinary bodies in the UK and elsewhere. But what is really astounding in the claims by Shorvon and his expert witness that his breaches were 'minor' and not worthy of disciplining is that they are all in clear contravention of the GMC's *own* guidance on good practice in research! Once again we see that disregard of patients' privacy, their right to refuse to participate, and even their safety is seen as above criticism, when some doctors are considering the ethical aspects of their colleagues' research.

These two cases illustrate why it is so important to follow up on the BAC's concern (shared by many bodies internationally) that the ethical hazards in research conducted by clinicians be fully recognised and measures taken

[11] Shorvon S. The Prosecution of Research — Experience from Singapore, *Lancet* **369**, 9576 (2007): 1835–1837.

to minimise them. We shall now look at three aspects in more detail: the potential confusion between clinical research and ordinary clinical care, a confusion which increases the tendency of patients enrolled in clinical trials to believe that they will get a therapeutic benefit; the ever-increasing COI faced by clinician-researchers; and the struggle to find adequate guidelines and educational methods to deal with these problems.

DISTINGUISHING BETWEEN TREATMENT AND RESEARCH

As we saw in the Shorvon case, at least some participants either were not informed or failed to understand that omitting their medication the night before reporting to the clinic for tests and being videotaped had nothing to do with their treatment, but was entirely for the purposes of a research project. Similarly, Green's patients were led to believe that the constant visits to the clinic and the repeated diagnostic procedures they were subjected to were part of a standard treatment regimen for their condition rather than an attempt to prove his hypothesis about CIS. It thus becomes essential ethically to make a clear distinction between treatment (whether experimental or otherwise) and clinical research, so that the interests of patients can be protected and properly informed consent for the research obtained.

There is an abundance of literature on research ethics that articulates how clinical research is distinct from standard medical practice. This distinction cannot be based on delineating different risk-benefit ratios that obtain as a matter of fact between the two activities.[12] It is not true that patients participating in clinical trials, by virtue of the nature and methods of research itself, would always have worse outcomes than they would if they received treatment in standard practice. Patient-subjects often receive regular and careful medical attention (typically more so than the case in standard medical care) from a research team that includes expert physicians and other healthcare professionals, and they gain access to new, effective, and safe interventions that might not be available outside the context of a trial. (As part of the benefits, some clinical trials provide all treatment and follow-up care at no cost to the patient.) Admittedly, a randomised clinical trial that uses placebo controls when a proven effective treatment exists presents a less favourable risk-benefit ratio than standard medical care, at least for those who are assigned

[12] Miller FG. Consent to Clinical Research, in *The Ethics of Consent: Theory and Practice*, eds. Miller FG and Wertheimer A. Oxford: Oxford University Press, 2010, pp. 375–404.

the placebo. But this is only a subset of clinical trials. A study on oncology trials showed that trial participation is often linked with higher patient survival rate for the cancers being studied.[13] In a systematic review of randomised clinical trials conducted across various medical areas, Vist *et al.* concluded that patients, when invited to participate in a randomised clinical trial, can be told that trial participation "is likely to result in similar outcomes to patients who receive the same or similar treatments outside a trial".[14]

Nor can clinical research be distinguished from standard practice based on the *uncertainty* of risks or benefits of interventions between the two activities.[15] Clearly, uncertainties of benefits or risks can permeate standard medical practice. For example, a physician may switch from one drug to another during the course of therapy to see what works best for a particular patient in line with that patient's goals (such as to avoid certain side effects). Physicians may also experiment with current interventions and techniques in the hope that innovation may benefit patients who have not been helped by standard methods. As Morreim points out, "standards of medical practice shift constantly as new information shows which accepted interventions are, and which are not, genuinely useful."[16]

What distinguishes clinical research from therapy is the difference in their defining goals, and thus what ought to be the professional orientation of those who are seeking to realise those goals. The defining goal of clinical research is to produce generalisable knowledge that can improve medical care for the benefit of future patients. Standard medical practice, on the other hand, even when experimental or innovative interventions are involved, aims at direct benefit to specific patients, with a focus solely on their best interests. To maximise these direct benefits, physicians are expected to administer 'personal care', to tailor treatment, or change the regimen completely if needed, to address the individual condition and needs of particular patients.[17] Treatment

[13] Stiller CA. Centralised Treatment, Entry to Trials and Survival, *British Journal of Cancer* **70**, 2 (1994): 352–362.
[14] Vist GE *et al.* Systematic Review to Determine Whether Participation in a Trial Influences Outcome, *British Medical Journal* **330**, 7501 (2005): 1175.
[15] Morreim EH. Medical Research Litigation and Malpractice Tort Doctrines: Courts on a Learning Curve, *Houston Journal of Health Law and Policy* **4**, 1 (2003): 1–86, pp. 14–15.
[16] *Ibid.* p. 14.
[17] Fried C. *Medical Experimentation: Personal Integrity and Social Policy.* New York: American Elsevier, 1974.

flexibility in the choice and delivery of treatment may be accommodated to some extent in a trial protocol. In some cancer trials, patient-subjects can choose to become directly involved in decisions of dose escalation.[18] Nevertheless, the principle of personal care and idiosyncratic treatment patterns remain largely incompatible with the production of generalisable knowledge, which needs to be systematic in order to benefit broader populations;[19] if a less rigorous scientific approach is used, false evaluations of novel treatments may harm future patients and place burdens on society and its healthcare system.

In addition, physicians are permitted to do harm to the patient only if there is some compensating benefit that directly accrues to the patient. But physicians as researchers may administer risky or painful procedures, such as lumbar punctures, solely to measure trial outcomes or to gather other research data. Patients can benefit from the data collected from these procedures, but this is a matter of incidental benefit rather than design. Unlike the physician-patient relationship, the researcher-subject relationship cannot be viewed as fiduciary in nature.[20] From a legal perspective, fiduciaries are expected to exercise their discretion and authority solely in the entrustor's interests, which are to be prioritised over the interests of third parties. In clinical research, investigators are expected to subordinate or fit the therapeutic goals of patients to scientific and social goals. It is for this reason that getting the balance right as a clinician-researcher is so difficult, and why the fully informed consent of patient-subjects is essential.

To be sure, some trial protocols have clinical benefit for patient-subjects as an endpoint (but not the only endpoint). What is problematic is that sometimes, when 'clinical benefit' is indicated as an endpoint, it does not correlate with any benefit that subjects seek or even experience. For example, Ohorodnyk and colleagues found that the majority of oncology trials use the term 'clinical benefit' to describe the endpoint of objective tumour findings (such as a decrease in tumour size), rather than an improved quantity

[18]Daugherty CK *et al.* Study of Cohort-Specific Consent and Patient Control in Phase I Cancer Trials, *Journal of Clinical Oncology* **16**, 7 (1998): 2305–2312.
[19]King NMP. Defining and Describing Benefit Appropriately in Clinical Trials, *Journal of Law and Medical Ethics* **28**, 4 (2000): 332–343.
[20]Morreim EH. The Clinical Investigator as Fiduciary: Discarding a Misguided Idea, *Journal of Law and Medical Ethics* **33**, 3 (2005): 586–598.

and/or quality of life for the subjects.[21] As they point out, using the term in this way is inconsistent with a patient-centric definition (as it was originally intended in the context of oncology trials) and may, as a result, lead patients to misinterpret the benefits of research participation.

The Therapeutic Misconception

This leads to the more general problem of the tendency of patients to believe that clinical research is bound to be to their individual benefit. It may well be that patient-subjects, or at least certain groups, will invariably conflate the scientific orientation of clinical research with the therapeutic orientation of standard medical care even when technical terms and scientific methods are explained and made understandable to them. Indeed, studying the informed consent of psychiatric patients participating in randomised clinical trials, Appelbaum *et al.* — who coined the term 'therapeutic misconception' in 1982 to describe patient-subjects' tendency to conflate clinical research with practice — found that patient-subjects failed to understand or appreciate that their trial participation would not operate within a therapeutic, patient-centric framework even when they could conceptually grasp key aspects of trial design.[22] For example, some patients who knew the meaning of randomisation did not think that this procedure would ultimately be applied in the study.

In a recent study led by Appelbaum on the frequency of the therapeutic misconception, it was found that 62% of 225 patient-subjects in 44 clinical trials covering a wide range of medical conditions harboured the therapeutic misconception.[23] Patient-subjects undergoing trials to treat non-life-threatening illnesses like plantar warts were also found to be under the influence of the misconception. (Remarkably, the therapeutic misconception has also been observed in patient-subjects participating in Phase I oncology trials, despite

[21] Ohorodnyk P, Eisenhauer EA and Booth CM. Clinical Benefit in Oncology Trials: Is This a Patient-Centred or Tumour-Centred End-point? *European Journal of Cancer* **45**, 13 (2009): 224–225.

[22] Appelbaum PS, Roth LH and Lidz CW. The Therapeutic Misconception: Informed Consent in Psychiatric Research, *International Journal of Law and Psychiatry* **5**, 3–4 (1982): 319–329; and Appelbaum PS *et al.* False Hopes and Best Data: Consent to Research and the Therapeutic Misconception, *Hastings Center Report* **17**, 2 (1987): 20–24.

[23] Appelbaum PS, Lidz CW and Grisso T. Therapeutic Misconception in Clinical Research: Frequency and Risk Factors, *IRB* **26**, 2 (2004): 1–8.

the fact that such trials are traditionally designed to find out about dose levels and the associated probabilities of toxicity of an investigational drug.[24] Historically, such trials present a low clinical improvement rate of around 5%.) The phenomenon does not seem to be associated with any detected cognitive impairment or loss of reality in grasping at treatment straws. Without doubt, as empirical studies show, the therapeutic misconception is prevalent among all kinds of patient-subjects. One can surmise that it manifests easily in patients' mindset because of their dependent, vulnerable state and, arguably, because of a misplaced trust in physicians wearing the hat of researchers.[25]

The Question of Consent

The first principle of the Nuremberg Code is that "the voluntary consent of the human subject is absolutely essential" and this requires the subject to adequately comprehend trial elements such that the subject is able to "make an understanding and enlightened decision".[26] Clearly, the therapeutic misconception hinders perfect understanding and a fully rational decision to participate. Does the mere existence of the therapeutic misconception in patient-subjects invalidate their consent and warrant their exclusion from trial participation?

Many commentators would argue against this conclusion. As Sreenivasan points out, it is a mistake to confuse an "ethical aspiration with a minimum ethical standard".[27] Fully informed consent is an ideal rarely (if ever) achieved in clinical settings. Merz and Fischhoff warn us that the notion of consent would turn out to be a legal fiction if we expect all patients "to make informed, knowledgeable, rational decisions in their own best interest".[28] Indeed, to insist on this idealised mode of decision-making could be argued as violating the autonomy of patients who may prefer to use less

[24]Glannon W. Phase I Oncology Trials: Why the Therapeutic Misconception Will Not Go Away, *Journal of Medical Ethics* **32**, 5 (2006): 252–255.
[25]de Melo-Martín I and Ho A. Beyond Informed Consent: The Therapeutic Misconception and Trust, *Journal of Medical Ethics* **34**, 3 (2008): 202–205.
[26]*Trials of War Criminals before the Nuremberg Military Tribunals under Control Council Law*, No 10, Vol 2. Washington, DC: US Government Printing Office, 1949, pp. 181–182.
[27]Sreenivasan G. Does Informed Consent to Research Require Comprehension? *Lancet* **362**, 9400 (2003): 2016–2018.
[28]Merz JF and Fischhoff B. Informed Consent Does Not Mean Rational Consent: Cognitive Limitations on Decision-Making, *Journal of Legal Medicine* **11**, 3 (1990): 321–350, p. 322.

rational ways of making decisions.[29] Patient-subjects may choose to participate in a trial, even if the odds of benefiting clinically are less than 'a roll of the dice', because of optimism or hope, which can contribute positively to the therapeutic (or perhaps palliative) process. Excluding prospective subjects from research participation on the count that their decision-making is influenced by some misconception may cause unjustified harm to them.

Importantly, as Miller argues, we should recognise that the therapeutic misconception is a matter of degree.[30] The significance of the misconception for invalidating consent can be assessed only by knowing the particular aspects of particular trials, including design, the procedures used and for what purpose, and the type of treatments being tested. That a patient may be confused about some aspects of a particular trial may not be as ethically worrying as the case of another patient confused about some other aspects. It has been suggested as a rule of thumb that the therapeutic misconception should be tolerated when there is little likelihood that the individual would have made a different decision in the absence of the misconception.[31] To be more specific, the therapeutic misconception does not compromise autonomy when the design of a trial is procedurally similar to standard medical care; when no risky or burdensome procedures are administered solely for research purposes; and when there is a high probability of or exclusive chance (for example, when the intervention is available only in a trial setting) for clinical benefits with low risk of harm involved.[32] Some Phase III trials satisfy these conditions. It is unlikely that patient-subjects would have made a different decision in the absence of the therapeutic misconception in trials that satisfy all three conditions.

While the therapeutic misconception may be ethically tolerable in some circumstances, this does not exempt researchers and other stakeholders from the ethical obligation to do their best to dispel the therapeutic misconception

---

[29] Miller FG. Consent to Clinical Research, in *The Ethics of Consent: Theory and Practice*, eds. Miller FG and Wertheimer A. Oxford: Oxford University Press, 2010, pp. 375–404.
[30] *Ibid.* p. 389.
[31] Miller FG and Joffe S. Evaluating the Therapeutic Misconception, *Kennedy Institute of Ethics Journal* **16**, 4 (2006): 353–366.
[32] Voo TC. Therapeutic Misconception, in *International Encyclopaedia of Applied Ethics*, eds. Chadwick R, Callahan D and Singer P. London: Academic Press, forthcoming in 2010.

in prospective and current patient-subjects.[33] While it seems plausible that the therapeutic misconception may persist in some patient-subjects despite the best of efforts to dispel it, certain efforts to overcome it, such as the use of neutral educators to impress upon prospective patient-subjects how trial participation differs from standard medical care, have been shown to substantially improve subjects' understanding of key aspects of trial design.[34] It is always ethically preferable that patients provide consent for research participation without being influenced by the therapeutic misconception. Even for trial participation with a risk-benefit ratio no less favourable than that of standard medical care, patient-subjects should grasp the distinction between research and practice to understand in a meaningful way what they have chosen and to appreciate what they have forgone, with their consent to participation, such as treatment alternatives to participation in the trial, time, and personal convenience. There is always an element of sacrifice in choosing to help in medical research.

DEALING WITH CONFLICTS OF INTEREST

A still more serious ethical challenge is faced by clinician-researchers working in today's medical research environment. This is the pressure to recruit patients and to produce research results.

Apart from medical research in academic settings of the kind described in the cases of Dr Green and Professor Shorvon, acceleration of the research process in the 1980s and 1990s by for-profit companies has augmented the problems associated with medical research in clinical practice. In 2002, the American Medical Association Council on Ethical and Judicial Affairs published an article in its journal by Morin *et al.* which noted that profitability from increased investment in research and development meant that leading pharmaceutical companies had to aim at launching between 24 and 34 new drugs each year.[35] To achieve cost efficiency and bypass cumbersome review

[33] Horng S and Grady C. Misunderstanding in Clinical Research: Distinguishing Therapeutic Misconception, Therapeutic Misestimation, and Therapeutic Optimism, *IRB* **25**, 1 (2003): 11–16.
[34] Flory J and Emanuel E. Interventions to Improve Research Participants' Understanding in Informed Consent for Research: A Systematic Review, *Journal of the American Medical Association* **292**, 13 (2004): 1593–1601.
[35] Morin K *et al.* Managing Conflicts of Interest in the Conduct of Clinical Trials, *Journal of the American Medical Association* **287**, 1 (2002): 78–84.

processes, industry now relies increasingly on contract research companies and site management organisations which enlist doctors in the private sector in research outside academic institutions. Morin *et al.* reported in 2002 that investigators based in academic medical centres decreased from 80% to 46% over a 10-year period; 60% of industry funding was allocated to community-based trials, representing a threefold increase in less than a decade. In academic medical centres, research physicians need scientifically reliable results and publications for professional advancement and future funding.

In 2000, at a US conference on 'Human Subject Protection and Financial Conflict of Interest' co-sponsored by the Department of Health and Human Services, National Institutes of Health (NIH), Food and Drug Administration (FDA), and Centers for Disease Control and Prevention, the director of the Office for Human Research Protection remarked to the 700 attendees that physician COI had "gotten entirely out of control". Subsequent to this event, the NIH issued interim guidance on financial relationships in clinical research, and a final document appeared in 2004.[36] In 2002, the Council on Ethical and Judicial Affairs of the American Medical Association (AMA) proposed a comprehensive set of measures published by Morin *et al.*, noting that high societal expectations that the burdens of disease and disability can be reduced through research, and increased research funding by private industry, have created a research imperative resistant to moderation. The measures included safeguards on subjects' welfare such as separating the roles of clinician and researcher by requiring non-treating healthcare professionals only to obtain patient consent and provide information; and ensuring the integrity of the informed consent process, which has to emphasise how treatment in research settings differs from ordinary treatment, the additional risks involved, and the lack of direct benefit to the participant.

In the area of safeguards against financial conflicts, the AMA proposal noted the relatively large sums in reimbursement of volunteers that were being paid by for-profit corporations for the purpose of rapid recruitment and unequivocally denounced as unethical finder's fees for mere referrals of patients for research (and not for medical service). It stressed the need to

[36] US Department of Health and Human Services, Office of Human Research Protection. Financial Relationships in Interests in Research Involving Human Subjects: Guidance for Human Subject Protection, 5 May 2004.

understand and identify the impact of financial incentives, so that potential conflicts may be "avoided, disclosed, or mitigated".[37] In addition to disclosure of COI to IRBs, it proposed disclosure statements on financial arrangements to other parties (such as the FDA, peer-reviewed journals, and potential research subjects) during and after completion of the trial to protect the integrity of scientific reports. Anti-kickback laws and anti-fraud measures would also serve to mitigate or prevent other potential conflicts, such as receiving grants from industry to perform studies of little scientific value and requiring no scientific research.

Perhaps most significantly, the AMA proposal concluded with a set of guiding norms addressed directly to clinician-researchers, recommending:[38]

1. Clinician participation in research only when they are satisfied that protocols are scientifically sound and within their areas of competence;
2. Clinician familiarity with research ethics and compliance with IRB review processes and government regulations;
3. Clinician responsibility for protecting participants from the effects of financial incentives through the informed consent process;
4. Clinician integrity in receiving compensation commensurate with research effort, rather than volume of subjects enrolled or referred;
5. Clinician responsibility for scrutiny of protocols and ensuring provisions for funding treatment in the event of complications;
6. Clinician honesty in disclosing to participants the nature and source of funding offered to investigators; and
7. Clinician responsibility for declining to participate in studies where there is no assurance that publication of results will not be delayed or obstructed by sponsors.

The Inadequacy of Conflicts of Interest Guidelines Alone

This last set of recommendations underscores a belief that without clinician integrity within research settings, regulations, however prolific, will not eliminate COI that continue to characterise the clinician-researcher's predicament in the face of the unbridled research imperative and market-oriented research

[37] Morin K *et al.* Managing Conflicts of Interest in the Conduct of Clinical Trials, *Journal of the American Medical Association* **287**, 1 (2002): 78–84, p. 81.
[38] *Ibid.* pp. 83–84.

practices. However, banking on individual integrity alone is not likely to hold out any great hope of change either.

It is a commonplace that regulatory guidance is only as effective as its commitment to monitoring and pinpointing responsibility for compliance. Even with guidance that identifies for clinician-researchers certain norms of responsible behaviour, such as making financial disclosures and receiving only due compensation, there is now overwhelming evidence in social science research that individuals are not always conscious of their motives and often fail to act in ways they believe they would.[39] Expectations of reciprocity operate in powerful and subtle ways that have negative consequences for clinical care.[40] For example, the rate of drug prescriptions by clinicians or requests for additions to hospital drug formularies has been found to be positively correlated with receiving gifts (even small gifts) from drug company representatives, accepting samples, and attending company-supported symposia — a clear but subtle deviation from the norms of evidence-based medical practice.[41] Disclosure may sanitise COI and lead clinicians to pretend they have disappeared. Or the effort to avoid disclosure requirements may lead clinicians to be 'creative' in their interpretation of what is or is not a COI.[42]

Medical Ethics in Academic Medicine

For these reasons, there have been calls for academic medical centres, rather than individual clinicians, to lead efforts to eliminate COI in the current climate of undue influence on medical objectivity. Brennan *et al.* have identified a clear obligation that academic medicine has in this regard, arguing that the profession and industry look to academic medical centres for influential

[39] Dana J and Loewenstein G. A Social Science Perspective on Gifts to Physicians From Industry, *Journal of the American Medical Association* **290**, 2 (2003): 252–255.
[40] Wazana A. Physicians and the Pharmaceutical Industry: Is a Gift Ever Just a Gift? *Journal of the American Medical Association* **283**, 3 (2000): 373–380.
[41] Chren MM and Landefeld CS. Physicians' Behavior and Their Interactions With Drug Companies: A Controlled Study of Physicians Who Requested Additions to a Hospital Drug Formulary, *Journal of the American Medical Association* **271**, 9 (1994): 684–689; Cleary JD. Impact of Pharmaceutical Sales Representatives on Physician Antibiotic Prescribing, *Journal of Pharmacy Technology* **8** (1992): 27–29; and Peay MY and Peay ER. The Role of Commercial Sources in the Adoption of a New Drug, *Social Science and Medicine* **26**, 12 (1988): 1183–1189.
[42] Cain DM, Loewenstein G and Moore DA. The Dirt on Coming Clean: Possible Effects of Disclosing Conflicts of Interest, *Journal of Legal Studies* **34**, 1 (2005): 1–24.

and independent advice; that independent research into the impact of new drugs and devices on population health is borne mainly by them; and that they have primary responsibility for training future clinicians in the habits of scientific objectivity and integrity.[43]

According to a report, Harvard Medical School is the only academic institution placing absolute limits on economic COI between its researchers and the human subjects of clinical trials. Members of the Harvard medical faculty may not conduct research in academic laboratories or teaching hospitals for companies in which they hold more than $20,000 worth of stock, or from which they receive more than $10,000 in consulting fees or royalties.[44]

In a study of COI policies at the ten US medical schools receiving the highest amounts of NIH research funds, Lo *et al.* found that, in comparison with medical schools, stricter COI policies prevailed in some industry-sponsored clinical trials. In addition to prohibiting investigators from holding stock or stock options in the company whose product is being studied, honorariums for speaking engagements, payments for consulting, and reimbursements of any kind from corporate sponsors were forbidden within a year of the publication of research findings. The authors suggest that university-based investigators, the research team, and immediate families be prohibited from holding stock, stock options, or decision-making positions in the sponsoring companies, companies that manufacture the product or device being tested, and companies that manufacture competing products and devices. They would not be prevented, however, from having grants or contracts from such companies that support their research.[45]

THE FUTURE: CAN WE OVERCOME THE ETHICAL HAZARDS?

The title of our paper derives from the saying in the New Testament that no one can be the servant of two masters, with the warning that "you cannot

[43] Brennan TA *et al.* Health Industry Practices That Create Conflicts of Interest: A Policy Proposal for Academic Medical Centers, *Journal of the American Medical Association* **295**, 4 (2006): 429–433.

[44] Miller FH. Trusting Doctors: Tricky Business When It Comes to Clinical Research, *Boston University School of Law Working Paper Series.* Public Law & Legal Theory: Working paper no. 01-09.

[45] Lo B, Wolf LE and Berkeley A. Conflict-of-Interest Policies for Investigators in Clinical Trials, *New England Journal of Medicine* **343**, 22 (2000): 1616–1620.

serve both God and mammon".[46] Perhaps the reference is apposite, given the temptations evident in some areas of clinical research. But even if financial COI does not come into the picture, clinicians, with the worthy aim of improving medical knowledge for the benefit of all patients in the future, may fail to perceive some of the hazards in having two potentially conflicting aims. The purpose of our chapter has been to highlight this problem and to encourage more transparency on the part of the institutions encouraging research regarding the ethical hazards. The reality is that the two aims cannot always be reconciled (for example, when a drug trial has a placebo arm), and it is only honest to admit this to patients rather than to rely on the therapeutic misconception to boost enrolment. By moving from the term of 'experimental subject' to 'research participant', we can begin to treat patients with the respect they deserve, as people who also would like to assist in improving medical care for all. In this context it is the language of partnership which seems the most appropriate — a sharp contrast with the paternalistic notion that patients are too anxious or too stupid to understand the nature of a clinical trial.

If this greater degree of openness and genuine partnership is the aim, then — although the conflict cannot be overcome — we can suggest practical ways to make it more ethically acceptable. First, there has to be a concerted effort to raise the standard of ethical awareness in the medical research community. Too often researchers see ethical appraisal as something external to themselves and as a barrier to pursuing their research rather than an aid to doing it better. This externalisation of ethical responsibility may well have been fostered by the extensive array of ethical review bodies now in place, each with their own bureaucratic procedures. It is hardly surprising that researchers come to see ethical approval as merely a hurdle to be jumped rather than an opportunity to think through the acceptability of their proposal. The solution to this problem has to be better education in both research ethics and research integrity across the *whole* spectrum of medical research, with courses being made mandatory not just for doctors and other healthcare professionals but also for their scientific colleagues.

Second, although policies alone cannot produce ethical behaviour in the research community, we believe that all academic medical centres need to spell out quite explicit policies in this area, especially where COI arise, and

[46]The Bible, American King James Version, *Matthew* 6:24.

such policies must be fully open to external scrutiny to make sure that they conform to the best international standards available. Gaps in policies are likely to be identified in the expansion of research and collaboration, and researchers need to engage in regular dialogue and feedback with regulatory agencies to generate more proactive, rather than reactive, solutions to ethical and legal dilemmas.

Last, it seems to us that after ten years of existence, and turning to the future, a likely next task for the BAC will arise from the increasing need to consider the ethics of translational clinical research. The separation of clinical and biomedical ethics will become increasingly difficult to sustain. The catchphrase "from bench to bedside", now so often used to describe the new emphasis on translational clinical research, makes it obvious that an advisory body concerned with the ethics of biomedical research cannot allow a separation to be made between clinical ethics and biomedical ethics. The two go hand in hand, and scientists 'at the bench' bear just as much responsibility as their clinical collaborators for the effects of their research on individual patients and on society as a whole. The notion of a value-free science is unsustainable, and so the collaboration is not just scientific — it is also ethical or (hopefully) at least not unethical. As Singapore increasingly becomes an international powerhouse for biomedical and clinical research, a fresh report on the ethics of such research could be of great national and international importance in the next decade.

6

The US Model for Oversight of Human Stem Cell Research

Lindsay Parham and Bernard Lo

Oversight of human stem cell research (hSCR) is essential to ensure that this research is carried out in an ethically responsible manner and to maintain public trust. The US model for such research illustrates several important policy issues in research oversight. In the US oversight occurs at multiple levels, often resulting in inconsistent requirements and can lead to ethical dilemmas and administrative burdens. This complex system of oversight has been shaped by American culture, and while it may be difficult for scientists, government officials, and policy scholars in other countries to understand, it is nevertheless consistent with American attitudes toward government regulation. The US oversight system has evolved over time, and has been influenced by research on stem cell policy and developments in stem cell science. The US experience also illustrates policy issues that any system of research oversight must address. To begin with, however, we will provide some background on how stem cell research oversight has developed in the US.

OVERSIGHT OF HUMAN STEM CELL RESEARCH IN THE US

More so than countries such as Singapore, the US has multiple levels of research oversight, which can lead to inconsistency, confusion, and administrative burdens.

Multiple Levels of Oversight

Research oversight takes place through federal laws and regulations that are binding nationwide, state laws and regulations that are unique to each state, institutional policies that are specific to research institutions such as universities, and the policies of research sponsors. The scope of coverage of laws and regulations may vary; for example, regulations may cover all research that takes place in an institution or a state, or they may apply only to research funded by a particular sponsor. Some requirements apply to all human pluripotent stem cell lines (hSC lines), while others apply only to human embryonic stem cell lines (hESC lines). Laws and regulations are legally binding, while some policies are merely recommendations.

The US federal regulations for the *Protection of Human Subjects,*[1] which were first issued in 1974, are intended to ensure that research is conducted ethically and that research participants are protected. These regulations are technical and complex. Subpart A of the regulations, known as the Common Rule, has been adopted by 17 other federal agencies, including the Department of Health and Human Services, which encompasses the National Institutes of Health (NIH) and the Centers for Disease Control and Prevention. The Common Rule applies to the donation of human biological materials for stem cell research and to clinical trials involving stem cells.

The Common Rule does not apply to all research involving human participants, but only to such research that:

1. is funded by or carried out by federal agencies that have adopted the Common Rule;
2. is conducted by institutions that have promised that all research on human subjects will comply with the Common Rule, including research privately funded or conducted off-site; or
3. will be submitted to the Food and Drug Administration (FDA) in support of an investigational new drug (IND) or device application.

The FDA requires that any company wishing to market an unlicensed drug or product obtain its approval for an IND application before it may carry out clinical trials in humans. Furthermore, any company wishing to market a stem cell therapy has to obtain FDA approval first. For example, in August 2008 a company called Regenexx, which was advertising a therapy using stem cells

[1]Title 45, *US Code of Federal Regulations,* Part 46, Subparts A–D.

that had not been approved by the FDA, was warned by the FDA to comply with its policies or risk legal action.[2]

There are additional federal laws and regulations that apply specifically to stem cell research. Under the Dickey–Wicker Amendment,[3] researchers cannot use federal funds for research in which human embryos are created, destroyed, or discarded.

The President may issue Executive Orders, which do not require approval by Congress. In 2001, then President George Bush issued an Executive Order that allowed federal funding for embryonic stem cell research only for work with a small set of existing hESC lines that had been derived prior to the signing date of the executive order. In 2009, President Barack Obama issued an Executive Order that allowed federal funding for research with existing hESC lines derived from embryos that were created by *in vitro* fertilisation (IVF) for reproductive purposes but were no longer needed for that purpose.[4] The donation of embryos for such derivation must satisfy certain requirements for informed consent. However, the Dickey–Wicker Amendment continues to prohibit the derivation of new hESC lines with federal funds.

US state laws and regulations add an additional layer of complexity to the American regulatory framework. State laws may prescribe what research is permissible to be performed within the state or what research may be funded with state money. State laws on the issue vary widely; laws may restrict the use of embryonic stem cells for some or all sources, or specifically permit only certain activities. For example, as of 2008, California, Connecticut, Illinois, Iowa, Maryland, Massachusetts, New Jersey, and New York all had passed legislation that encourages stem cell research. Louisiana, on the other hand, strictly forbids research on IVF embryos. Illinois and Michigan prohibit research on live embryos. Arkansas, Indiana, Michigan, North Dakota, and South Dakota

[2] FDA Warns Regenerative Sciences About Unlicensed Drug. *FDA News*, 22 August 2008. Available at: http://www.fdanews.com/newsletter/article?articleId=109705&issueId=11887. See also Malarkey MA and the US Food and Drug Administration, Letter to Regenerative Sciences Inc, 25 July 2008.
[3] Section 128 of US Public Law 104–199.
[4] President Barack Obama — Executive Order B505 — Removing Barriers to Responsible Scientific Research Involving Human Stem Cells. 2009. Available at: http://edocket.access.gpo.gov/2009/pdf/E9-5441.pdf.

prohibit research on cloned embryos. Missouri forbids the use of state funds for reproductive cloning but not for cloning for research.[5]

Human stem cell research is also subject to the policies of institutions that carry out such research. These internal policies often cover what kinds of research can take place within the institution. Policies and guidelines may differ from one institution to the next; for example, using a hSC line derived from a gamete donor without informed consent would not be permissible for use at the University of California, San Francisco (UCSF), but might be permitted at another research institution.

Finally, research sponsors may set conditions for hSCR carried out under its funding. Sponsors range from governmental agencies such as the NIH to private foundations and commercial companies. Sponsors may have their own standards for research they fund. For example, stem cell research funded by the state of California must satisfy specific requirements regarding informed consent, compensation for oocyte donors, and types of research permitted. As another example, the Juvenile Diabetes Research Foundation requires hSCR to undergo review by a panel of national ethics experts.

Voluntary guidelines regarding stem cell research may be influential. In 2005, the US National Academy of Sciences (NAS) issued recommendations regarding human embryonic stem cell research (hESCR).[6] Because these recommendations were developed by a panel of experts and subjected to peer review, they have been widely adopted by research institutions and research funders. In addition, the International Society for Stem Cell Research (ISSCR) has issued voluntary guidelines[7] developed by an international panel.

CONSEQUENCES OF MULTIPLE LEVELS OF OVERSIGHT

Conflicting Oversight Requirements

Not surprisingly, this welter of laws, regulations, institutional policies, and voluntary guidelines may result in inconsistencies and confusion. Problems

[5] National Conference of State Legislatures, *Stem Cell Research*, Updated January 2008. Available at: http://www.ncsl.org/issuesresearch/health/embryonicandfetalresearchlaws/tabid/14413/default.aspx.

[6] National Research Council and Institute of Medicine. *Guidelines for Human Embryonic Stem Cell Research*. Washington, DC: National Academies Press, 2005.

[7] International Society for Stem Cell Research, *Guidelines for the Conduct of Human Embryonic Stem Cell Research*, 21 December 2006.

arise when a researcher at one institution wishes to use lines that were derived at another institution in a manner that would not have been permitted at the first institution. In response to such inconsistencies, some institutions have tried to develop policies that are consistent with different legal requirements. For example, the California Institute for Regenerative Medicine (CIRM) oversees the use of California state funds for hSCR. These regulations were designed to be consistent with the NAS guidelines, though they are stricter in some areas. Although CIRM regulations only cover the research it funds, many California universities and research institutions apply them to all stem cell research at their site. Hence policies may have much more influence than their technical legal scope.

As we have described, the US has a complex system of oversight that is not always internally consistent. For example, under the Common Rule, existing human biological materials may be anonymised and used for research without consent from the donors. Under the Common Rule it is therefore legally permissible for researchers to try to derive new embryonic stem cell lines, without obtaining permission, from frozen embryos remaining after a woman or couple has completed their infertility treatment. This might occur, for example, if an IVF practice received no response from IVF patients after asking their preference for disposition of frozen embryos. However, UCSF decided that because IVF patients vary in their preferences regarding the disposition of frozen embryos and that some are known to object to stem cell research, using frozen embryos for stem cell research without their consent would fail to respect the IVF patients as persons:[8]

> *...people commonly place great emotional and moral significance on their reproductive materials. Using gametes or embryos for certain kinds of research without consent, even after identifiers have been removed, could be regarded as wrong or offensive... We suggest that gamete donors' wishes should be determined and respected; informed consent from both oocyte and sperm donors should be obtained for an embryo to be used in research.*

Subsequently, the NAS, other institutions, and the California-funded stem cell program also adopted this requirement of specific consent for stem cell research.

[8] Lo B *et al.* Consent from Donors for Embryo and Stem Cell Research, *Science* **301** (2003): 921.

Ethical Dilemmas Resulting from Conflicting Requirements

Diverse and often conflicting policies on hESCR in the US may create ethical dilemmas, for example when a hESC line is being imported from another jurisdiction whose derivation would not be permitted at the importing institution. Even if it may not be illegal for researchers to import and work with such hSC lines, there may be serious concerns about undermining ethical standards.[9] For example, in 2007, the UK decided to allow women providing oocytes for research to receive some payments for lost wages or discounts on their IVF care.[10] Under such arrangements, UK researchers are now trying to derive a hSC line using somatic cell nuclear transfer. If their efforts are successful, other scientists around the world will want to carry out additional research with this hSC line. However, under the NAS guidelines for hSC research or under laws in states such as California and Massachusetts, donors of materials for stem cell research may not receive payments or other consideration in excess of out-of-pocket expenses.[11] Therefore, institutions in these jurisdictions will have to decide if their researchers can use lines derived under such circumstances.[12]

Administrative and Regulatory Burdens

Multiple levels of oversight of hSCR can lead to significant administrative and regulatory burdens, which decrease the pace of scientific discovery. At a meeting of the Interstate Alliance on Stem Cell Research in 2008, several persons cited confusing or contradictory state regulations and legislation as a cause for research delays and uncertainty. In Massachusetts, concerns have

[9]See Lo B *et al.* Importing Human Pluripotent Stem Cell Lines Derived at Another Institution: Tailoring Review to Ethical Concerns, *Cell Stem Cell* **4** (2009): 115–123; Daley GQ *et al.* The ISSCR Guidelines for Human Embryonic Stem Cell Research, *Science* **315** (2007): 603–604; Mathews DJ *et al.* Integrity in International Stem Cell Research Collaborations, *Science* **313** (2006): 921–922; and Skene L. Undertaking Research in Other Countries: National Ethico-Legal Barometers and International Ethical Consensus Statements, *Public Library of Science (PLoS) Medicine* **4** (2007): 0243–0247.

[10]Human Fertilisation & Embryology Authority, UK, FAQs for Donors, Updated 14 April 2009. Available at: http://www.hfea.gov.uk/2627.html.

[11]National Research Council and Institute of Medicine. *Guidelines for Human Embryonic Stem Cell Research.* Washington, DC: National Academies Press, 2005.

[12]Lo B *et al.* Importing Human Pluripotent Stem Cell Lines Derived at Another Institution: Tailoring Review to Ethical Concerns, *Cell Stem Cell* **4** (2009): 115–123.

been raised that regulatory complexity has created so many problems in research that researchers might leave the state to work elsewhere.[13]

In our experience at UCSF, it is difficult and time-consuming to obtain information on cell line provenance when wishing to use a hSC line derived in a different jurisdiction. Many institutions do not have readily available documentation of approval by an institutional review board (IRB) or copies of consent forms from embryo and gamete donors, or are not used to responding to requests for such information. Similarly, it may be difficult to obtain information about the review committee and process at another institution. It may be particularly difficult to obtain information about oversight for hSC lines derived by for-profit companies, especially if the company that derived the line was bought out by another. Currently there is no central repository that provides information on the oversight of the derivation of existing hSC stem cell lines, for example copies of the informed consent forms and documentation of IRB approval. Hence, each research institution must gather information on the oversight of the derivation of a particular hSC line. Such duplication of effort adds considerably to the administrative costs and burdens of research.[14]

THE IMPACT OF US CULTURE ON STEM CELL POLICY

A country's laws and regulations reflect its cultural values. This is particularly true for the US, a multi-cultural society in which disagreements over cultural and religious values have been prominent. A prime example of such a disagreement is the US anti-abortion movement, commonly referred to as the 'pro-life' or 'right-to-life' movement. Opposition to hESC research in the US is often associated with opposition to abortion and with the pro-life movement. Many Americans believe that a human embryo has the moral status of a person, just like a live-born child or an adult. In this view, human life 'begins at conception' and an embryo is thus a person and should be accorded rights and legal protections as such.[15] Opinion polls show that Americans are deeply divided about reproductive rights and that views on

[13] Interstate Alliance on Stem Cell Research, Meeting Summary, 9–10 April 2008. Available at: http://iascr.org/docs/MeetingSummary-Apr2008.pdf.
[14] Lo B *et al.* Importing Human Pluripotent Stem Cell Lines Derived at Another Institution: Tailoring Review to Ethical Concerns, *Cell Stem Cell* **4** (2009): 115–123.
[15] Lo B and Parham L. Ethical Issues in Stem Cell Research, *Endocrine Reviews* **30**, 3 (2009): 204–213.

abortion have remained remarkably unchanged over the years. According to the Pew Research Center, abortion "continues to split the country down the middle".[16] When asked their opinion on abortion, around 35% of respondents in a 2006 poll felt abortion should be illegal with few exceptions, 11% felt abortion should be banned in all circumstances, including cases of rape, incest, and threat to the mother's life, 20% felt abortion should be legal but under stricter regulation, and 31% preferred abortion to be generally available.[17]

Such pro-life and right-to-life beliefs have had significant influence on public policy towards hESC research. Under the Bush administration, NIH funds for stem cell research were heavily restricted, and programs for 'adopting' frozen IVF embryos were promoted as opposed to donation to scientific research. However, such opposition to stem cell research is not monolithic and has evolved over the years. Several conservative and pro-life leaders now support stem cell research using frozen embryos remaining after a woman or couple has completed infertility treatment and has chosen to donate to research as opposed to another couple.[18]

Examples of this evolution in hESC research support among pro-life advocates are former First Lady Nancy Reagan and US Senator Orrin Hatch of Utah. In 2001, Senator Hatch and Senator Bill Frist called for a bill restricting federal funding for stem cell research,[19] a more restrictive position than President Bush's Executive Order. Senator Hatch, a devout Mormon, declared, "I can't be for something that destroys human life".[20] By 2005, however, Senator Hatch had become a vocal supporter of embryonic stem cell research and supported a bill in the Senate to lift the restrictions on federal funding

[16]The Pew Research Center, *Most Want Middle Ground on Abortion: Pragmatic Americans Liberal and Conservative on Social Issues*, 3 August 2006.
[17]*Ibid.*
[18]Lo B and Parham L. Ethical Issues in Stem Cell Research, *Endocrine Reviews* **30**, 3 (2009): 204–213.
[19]Godoy M, Palca J and Novey B. Key Moments in the Stem Cell Debate, *National Public Radio*, 20 November 2007. Available at: http://www.npr.org/templates/story/ story.php?storyId=5252449.
[20]Senator Orrin Hatch Aborts Views to Support Stem Cell Research, *Knight Ridder Tribune*, 17 February 2003. Available at: http://media.www.clarksonintegrator.com/media/storage/paper280/news/2003/02/17/Features/Senator.Orrin.Hatch.Aborts.Views.To.Support.Stem.Cell.Research-372910.shtml.

put in place by President Bush in 2001. The Senator's views were stated on his website:[21]

> *The support of embryonic stem cell research is consistent with pro-life, pro-family values. ... I believe that human life begins in the womb, not a Petri dish or refrigerator. ... To me, the morality of the situation dictates that these embryos, which are routinely discarded, be used to improve and save lives. The tragedy would be in not using these embryos to save lives when the alternative is that they would be discarded.*

Americans' beliefs similarly have changed to support embryonic stem cell research.[22] A "clear majority" of 56% of those polled by Pew Research Center in 2006 supported research, up 13% from polls in the past.[23] In a 2008 Gallup poll on morality and beliefs, 62% of Americans said they believed hESCR was morally acceptable. A 2009 Gallup poll taken before President Obama lifted the Bush administration restrictions on funding shows that 52% of Americans believe the government should either ease restrictions or place no restrictions at all on hESCR. Thus although disagreements persist, it has been possible to forge public policy agreements on this highly contested topic.

In addition to religious beliefs, policies on hSCR in the US have been shaped by attitudes towards government. In the US, opposition to government regulation is widespread. Many people oppose 'big government' and view it as ineffective, inefficient, and an infringement on individual liberties. Many Americans believe in the 'free market', unfettered by government regulation, as the best means to economic prosperity. In this view, regulation thwarts individual initiative, which is commonly viewed as the engine of human flourishing. Thus, rigorous government regulation of research is often not trusted.

Because of these beliefs, models of stem cell research oversight that have been adopted in other countries would not be accepted in the US. In the

---

[21] Lo B and Parham L. Ethical Issues in Stem Cell Research, *Endocrine Reviews* **30**, 3 (2009): 204–213.

[22] The Pew Research Center. Vestal C, Embryonic Stem Cell Research Divides States, 21 June 2007.

[23] *Ibid.*

UK, the government licenses institutions that carry out stem cell research and carries out site visits and audits. In contrast, in the US, such an active role for government would be regarded by many as intrusive, ineffective, and wasteful. In Singapore, close partnerships between government, academia, and industry are generally believed to be the most efficient way to achieve social goals and economic growth. Government officials are respected as talented and beyond corruption, which is generally not believed to be the case in the US.

SOURCES OF REGULATORY STANDARDS AND BEST PRACTICES FOR OVERSIGHT

Ideas on stem cell policy that are not at first accepted may later have indirect but extensive influence on regulation and oversight. An example is the 1994 NIH human embryo research panel, whose recommendations to allow federal funding for certain types of human embryo research, including the derivation of hESC lines from embryos originally created for reproductive purposes, were not accepted.[24] While the panel had no influence on policy at the time, several recommendations in the report were widely adopted later: for example, a 14-day limit on the development of *in vitro* embryos being used in research, explicit consent from embryo donors, and restrictions on the use of federal funds for certain types of research.[25]

Local institutional policies can also influence public policy on a state or national level. An example of this kind of 'ground-up' development of standards is the widespread adoption of UCSF's policy requiring consent from gamete donors for the derivation of hESC lines. As discussed earlier, under the Common Rule human biological materials may be used in research without explicit consent from the donor after it has been anonymised. Because many people place great emotional and moral significance on their reproductive material, using such material without consent in research could be viewed as wrong or offensive.[26] For the derivation of embryonic stem cell lines from frozen IVF embryos, consent from the woman or couple in infertility treatment was readily accepted. Although consent from gamete donors

[24] National Institutes of Health, US. *Report of the Human Embryo Research Panel.* Bethesda, MD: National Institutes of Health, 1994.
[25] *Ibid.*
[26] Lo B *et al.* Consent from Donors for Embryo and Stem Cell Research, *Science* **301** (2003): 921.

was also recommended by the 1994 NIH panel, this idea was not implemented. In 2003, a US university carrying out cutting-edge embryonic stem cell research adopted as institutional policy a requirement for consent from any gamete donors and published the rationale for this policy in the journal *Science*. The policy explicitly rejected the use of existing reproductive materials without consent even if they were anonymised and urged a strong and broad view of respect for donors: "We suggest that gamete donors' wishes should be determined and respected; informed consent from both oocyte and sperm donors should be obtained for an embryo to be used in research".[27]

This policy of a leading stem cell research institution, published in a leading scientific journal, was widely adopted elsewhere. The NAS adopted this requirement for gamete donor consent in its influential 2005 *Guidelines for Human Embryonic Stem Cell Research*, citing this 2003 paper. While the NAS guidelines are recommendations, they have been adopted in several states and by many research institutions as legally binding. For example, the state of California adopted the NAS guidelines as interim regulations when it was still drafting its own regulations for stem cell research funding by the state. There are several reasons why the NAS guidelines have been so influential: an NAS panel is selected for balance and lack of bias, and its recommendations are evidence-based, consensus, and peer-reviewed.

The NAS guidelines also recommended that research institutions establish stem cell research oversight committees (SCROs) distinct from IRBs that oversee human subjects research. Most IRBs lack the time and scientific and ethical expertise required to oversee stem cell research. Furthermore, a great deal of stem cell research does not involve human subjects. For example, the injection of human pluripotent stem cells into non-human animals may not meet the regulatory definition of human subjects research if the stem cells have been anonymised. However, such injections may pose serious ethical concerns about the transmission of distinctively human characteristics or structures to non-human animals.

Because SCROs are novel, institutions had to develop SCRO policies and procedures. Such policies and procedures were first developed at institutions carrying out cutting-edge scientific research, then developed further through informal discussions among SCRO leaders at different institutions

[27] *Ibid.*

and through the peer-reviewed literature. Institutional policies, for example, may set requirements for membership and categories of research that must be reviewed or may not be approved.

Yet SCROs have faced a number of ethical issues that could not be anticipated when such institutional policies were first established. As an experienced SCRO, the UCSF committee wrote a series of papers describing their policies and procedures and the rationale for them. At one level, the UCSF SCRO explained how its policies and procedures worked on such issues as different levels of review, quorum, and conflicts of interest. At another level, the UCSF SCRO also developed an approach for the review of pluripotent stem cell lines derived at other institutions, addressing such issues as the use of older lines that had met ethical and legal standards in place at the time but not current standards, the level of documentation required regarding the original derivation of the line, and how to take into account specific scientific advantages of a line. These papers from an experienced SCRO are much more detailed and practical than the regulations or recommendations in the NAS report. Such ground-up development of best practices and their adoption as professional standards are an alternative to highly prescriptive regulations.

DIFFERENT MODELS OF RESEARCH OVERSIGHT

The US Approach: Specific Laws and Regulations

US laws and regulations on research are generally more specific than those in countries like Singapore. US regulations developed in the context of an adversarial legal system in a litigious society. In this setting, lack of specificity in laws and regulations is thought to increase lawsuits, because courts would be required to interpret regulations and specify the meaning of ambiguous terms. In the US there is also an emphasis on procedural justice and procedural safeguards. Procedural fairness is thought to require that those subject to laws and regulations — in this case researchers and research institutions — know in advance what exactly they must do to be in compliance. Lack of clear notice is regarded as unfair and also inefficient, because legal uncertainty will deter people from carrying out such important activities as biomedical research.

Laws and regulations in the US therefore are often very prescriptive. For example, Senate Bill Number 1260 (on standards for egg retrieval for stem cell research)[28] in California is a detailed law specifying what issues researchers must discuss with oocyte donors in California, particularly information about medical risks to donors. However, in a new and rapidly expanding field such as hSCR, laws and regulations need to be flexible to keep up with scientific developments and unforeseen situations. For instance, earlier discussions on hSCR regulation debated whether women who were donating oocytes specifically for hESCR should be compensated for their donated materials. Only later was the question brought up of payment to women providing oocytes from the same cycle for both infertility treatment and research.

Such 'oocyte sharing' might occur in several ways. First, some oocytes that were originally intended for infertility treatment may not be suitable for that purpose. For example, they may fail to fertilise or be too immature to be fertilised. Second, in rare situations, the woman or couple seeking IVF may not wish embryos that are not implanted to be frozen, so only a few oocytes retrieved are needed for infertility treatment. In these situations, any payments to oocyte providers were made in the context of IVF, where such payment is standard. If used for research, IVF payments are not considered an undue inducement to donate materials for research, because the donor is paid the same amount whether or not embryos are provided to researchers. Thus, even in states that forbid payment to women donating oocytes specifically for research, it seems reasonable to allow materials obtained in such a fashion to be used for research. This question was not considered in early discussions of payment, and would not be covered by laws and regulations that forbid payment for oocytes donated for research purposes.

There are additional concerns about oocyte sharing because the medical risks may be increased, for example, if physicians tried to increase the number of oocytes retrieved. An increased number of oocytes may be harvested through more aggressive hormonal manipulation, which in turn increases the risk of severe ovarian hyperstimulation syndrome. To address these concerns would require additional regulations or standards that restrict the intensity of hormonal manipulation and safeguard oocyte donors, for example by limiting the number of oocytes intended to be retrieved.[29] To further complicate

[28]Senate Bill Number 1260 became law on 1 January 2007, amending Health and Safety Code Section 125330 and Chapter 2 (commencing with Section 125330).
[29]Giudice L, Santa E and Pool R. *Assessing the Medical Risks of Human Oocyte Donation for Stem Cell Research.* Washington, DC: National Academies Press, 2007.

matters, in the US the clinical practice of IVF is not regulated by the government; for example, there are no legal restrictions on the number of oocytes that may be retrieved each cycle.

The 'Light Touch' Approach

The US approach to regulation has been criticised for overemphasising minor details of regulations, failing to address the major concerns about research ethics, and wasting resources on bureaucratic requirements. To critics, the IRB system focuses on documenting compliance with regulations rather than on substantive protections for human subjects. Furthermore, critics charge that draconian penalties are issued for minor infractions. As a consequence, valuable research may be stifled or delayed, without meaningful increases in protections for human subjects.[30]

In contrast to the very specific and detailed American regulations, some political scientists and governments favour more flexible oversight, commonly called responsive or 'light touch' regulatory theory. Responsive or light touch regulation has been characterised as "the use by regulators of mechanisms that are responsive to the context, conduct, and culture of those being regulated".[31] The emphasis is on flexible and goal-oriented regulations tailored to the most pressing problems, rather than on comprehensive rules and punitive sanctions: "One core concept is that where state intervention is required it should be the minimum necessary to achieve the desired outcome; consequently, it should target the highest risks and reward good practice with a light touch".[32]

Another feature of 'light touch' regulation is a decrease in the use of the "traditional 'command and control' mode of regulation, which focuses on regulation by the state through the use of legal rules backed by . . . sanctions".[33] In other words, light touch regulation calls on government to "steer, not row" when regulating and to create more "flexible, participatory and

[30] Fost N and Levine RJ. The Dysregulation of Human Subjects Research, *Journal of the American Medical Association* **298**, 18 (2007): 2196–2198.
[31] Healy J and Braithwaite J. Designing Safer Health Care Through Responsive Regulation, *Medical Journal of Australia* **184** (2006): S56–S59.
[32] Hogarth S. The Regulation of Nutrigenetic Testing: A Role for Civil Society Organisations? *Health Law Review*, 22 June 2008 (Special Edition).
[33] *Ibid.*

devolved forms of regulation".[34] An example of flexible regulation is the oversight of nursing homes and labour regulations, particularly in Australia.[35]

Arguments in Favour of 'Light Touch' Regulation

There are several arguments in favour of a 'light touch' approach. First, it uses scarce regulatory resources more efficiently. Because resources for oversight of research are limited, regulations should focus on the most serious problems rather than trying to cover all minor ethical concerns.

Second, a 'light touch' approach creates a culture of compliance that lessens the need for external oversight. This tries to build upon and reinforce standards of professional behaviour for researchers and physicians. Traditionally the medical and health professions are granted considerable scope for professional self-regulation,[36] with an understanding that doctors are dedicated to the best interests of patients and that researchers are committed to carrying out valid research. Standards of professionalism for researchers include ensuring an acceptable balance of benefits and risks and respecting research participants as persons. Professional standards are inculcated through education, mentoring, role modelling, and peer review.

The goal of creating a culture of compliance is not only to change behaviour, but also to alter the goals and values that motivate people. A culture of compliance creates social norms and expectations to act in ways that are consistent with an institution's policies or a profession's values. For instance, researchers are usually reluctant to face peer disapproval for violating professional norms. Ultimately these professional norms may be internalised as moral principles or virtues. Hence, in a culture of compliance, external sanctions and internal norms are aligned.

A third argument is that light touch regulation is effective. Oversight may be more effective and efficient if the institutions and individuals who are

[34] Healy J and Braithwaite J. Designing Safer Health Care Through Responsive Regulation, *Medical Journal of Australia* **184** (2006): S56–S59, p. S56.
[35] *Ibid.* In addition, see Howe J and Landau I. 'Light Touch' Labour Regulation by State Governments in Australia, *Melbourne University Law Review* **16** (2007). See also Braithwaite J. Regulating Nursing Homes: The Challenge of Regulating Care for Older People in Australia, *British Medical Journal* **323** (2001): 443–446.
[36] Healy J and Braithwaite J. Designing Safer Health Care Through Responsive Regulation, *Medical Journal of Australia* **184** (2006): S56–S59.

regulated play an active role in designing an oversight program that takes into account their needs and constraints. If such buy-in occurs, those who are subject to oversight are more likely to accept oversight standards as legitimate and to establish strong educational and mentoring programs to transmit them to younger researchers. On the other hand, if those who are regulated view the regulatory program as misguided, they may try to find loopholes, comply only minimally, or resist or evade requirements.

One particular aspect of light touch regulation that recently received great attention is the use of incentives. According to behavioural economics, incentives, 'nudges', and defaults lead people to act in desired ways. These techniques may achieve policy goals and be easier to implement than requiring or prohibiting specific behaviours. Tax incentives are the most common example. These techniques have not been widely applied in research oversight. Potential examples of incentives might be: in multi-centre studies, deferring to centralised review by national bodies; reduced documentation requirements for IRBs that have received accreditation; and taking institutional research oversight into account in applications for federal centre grants, program project grants, and training grants.

Fourth, flexibility is especially appropriate in areas that are rapidly changing. In stem cell research, new knowledge and techniques are constantly emerging. In such areas, detailed, rigid rules may quickly become obsolete. Obsolete rules that are no longer needed but are still in place may hinder the pace of research. Furthermore, flexibility allows rules to be adapted to cover new situations that could not have been foreseen when the oversight system was originally established.

Potential Problems with 'Light Touch' Regulation

A 'light touch' approach offers some potential problems as well as advantages.

First, highly specific regulations and severe punishments are sometimes necessary. Some behaviours are unacceptable and should be explicitly prohibited. One example is the failure to obtain informed consent from research participants or approval for a research project from an SCRO or an IRB. Another example is the breeding of non-human animals into which human pluripotent stem cells have been introduced, and the introduction of human pluripotent stem cells into human embryos or non-human primate embryos. For such serious infractions, sanctions should be strict and predictable.

Second, 'light touch' regulations may be difficult to defend after a scandal occurs. In the US, episodes of egregious misbehaviour have commonly evoked public outrage that nothing was done and have led to the enactment of highly specific regulations. After the Tuskegee scandal, during which treatment for syphilis was withheld from poor African-American participants in a research study, there was strong Congressional pressure to regulate federally funded research. Regulations were promulgated that later became the Common Rule.[37] In another example, Jesse Gelsinger, a volunteer in a Phase 1 gene transfer study, died as a result of participating in the study. Both the principal investigator in the trial and the university where the trial was conducted had a financial interest in the company whose product was being tested in the research.[38] There was strong Congressional and public pressure to respond to conflicts of interest in research, and the NIH enacted conflict of interest regulations for grantees.[39] As these episodes illustrate, after a fatal adverse event in research it is difficult to argue to the public and to Congress that the event was atypical and rare and that therefore additional regulations are not needed.

After a scandal occurs, 'light touch' regulations also may hamper efforts to punish offenders. Those accused of wrongdoing may argue that it was not made clear beforehand what they were required to do. Hence there was no specific rule that was violated. Thus rules that are not highly specific have the disadvantage of making noncompliance easier to explain or rationalise and making enforcement more open to challenge.

Lighter Touch Regulations in the US

In the 2004 election, California voters passed Proposition 71, which authorised $3 billion of public funding over ten years for stem cell research in California. With this ballot measure, California became the largest funder of stem cell research in the world. The measure established the California Institute

[37] Lo B. *Ethical Issues in Clinical Research: A Practical Guide*. Philadelphia, PA: Lippincott Williams & Wilkins, 2010, pp. 3–6.
[38] Steinbrook R. The Gelsinger Case, in *The Oxford Textbook of Clinical Research Ethics*, eds. Emanuel EJ *et al*. New York: Oxford University Press, 2008, pp. 110–120.
[39] *Ibid*.

for Regenerative Medicine (CIRM) to administer and oversee this research funding. CIRM developed regulations that have a lighter touch than many US regulations concerning research. Institutions receiving CIRM funding must establish a SCRO with appropriate expertise to review, approve, and oversee CIRM-funded stem cell research. CIRM gave institutions considerable flexibility in establishing SCROs and coordinating the review by SCROs, IRBs, and Institutional Animal Care and Use Committees.

The CIRM regulations furthermore try to avoid unnecessary regulatory burdens.[40] To avoid duplicative review, CIRM deemed some human stem cell lines acceptable for CIRM-funded research without further institutional review: stem cell lines approved by the US NIH, the UK Human Fertilisation and Embryology Authority, the UK Stem Cell Bank, or the Canadian Institutes of Health Research. The rationale is that a responsible body has already reviewed the derivation of these lines. Some types of hSCR — such as research with adult and cord blood stem cells and *in vitro* research with hESC — do not raise novel ethical concerns and therefore do not require in-depth review by each individual SCRO. However, in-depth SCRO review is required for research that raises complex ethical issues, including research involving oocytes and embryos and the introduction of human stem cells into humans and non-human animals. Thus, from a 'light touch' perspective, the oversight system should focus on the most serious problems.

CIRM recognised that collaboration among researchers from different institutions and countries would facilitate scientific progress. To encourage research cooperation, CIRM needed to address the fact that countries had different and sometimes inconsistent regulations regarding hESC research. CIRM did not insist on exact conformity to CIRM standards, but defined core requirements for hSC lines that CIRM-funded researchers might use. However, lines derived in other jurisdictions need not satisfy all the detailed requirements for researchers deriving hESC lines with CIRM funding. For example, imported hESC lines need not meet CIRM's heightened consent requirements for oocyte donors, although donors do need to provide informed and voluntary consent.

[40] Lomax GP, Hall ZH and Lo B. Responsible Oversight of Human Stem Cell Research: The California Institute for Regenerative Medicine's Medical and Ethical Standards, *Public Library of Science (PLoS) Medicine* **4**, 5 (2007): e114.

Because the public provided research funding for CIRM and has great interest in hESCR, public involvement in the formulation of the CIRM regulations was important. CIRM found various ways to engage the public during the process of drafting and revising the regulations.[41] First, CIRM meetings were public meetings at which the public had access to all written materials, such as briefing papers and draft guidelines, and were invited to comment on each topic discussed. These discussions often involved vigorous give-and-take discussions involving both a working group developing regulations and members of the public. The standard the working group developed was to respond to the concerns and objections raised by the public. When working group members did not agree with a suggestion, committee members were careful to explain the reasons for the decision. Second, CIRM held a series of meetings around the state at which the public was invited to offer input and comment on ethical issues related to stem cell research. The most frequent comments concerned the ethical use of human oocytes for nuclear transfer research. Finally, after the regulations were drafted, there was a 45-day period of formal public comment, during which CIRM responded in writing to all suggestions made. Many suggestions were incorporated into the final regulations.

Performance Rather than Prescriptive Standards

When CIRM established its regulations for hSCR, it favoured a lighter approach that acknowledged that the field was rapidly advancing and that there was little experience overseeing such research. The regulations set performance standards rather than prescriptive standards and encouraged SCRO committees to develop and adopt best practices.[42] For instance, California state laws specify that researchers must disclose certain information to women donating embryos for research. In contrast, CIRM emphasised comprehension by women donating oocytes for research rather than disclosure of information by researchers: donors need to comprehend the essential aspects of the research, including that the eggs will not be used for reproductive purposes and that there are medical risks associated with donation; that stem cell lines derived from donor oocytes may be grown in the lab and shared with other researchers; and that stem cell lines may be commercialised and patented but

---

[41] *Ibid.*
[42] *Ibid.*

donors will not share in patent rights or royalties.[43] The assessment of comprehension is not specified in the regulations, but rather left to researchers and IRBs to determine. These regulations work on the belief that, "with experience and evaluation, best practices for donor consent will be developed and broadly applied".[44] This process of developing and identifying best practices was felt to lead to a better process for consent than trying to specify at the onset through regulations how consent should be carried out.

Balancing Specificity and Flexibility

Regulations need to balance specificity and flexibility when overseeing a new and rapidly developing field of research such as hSCR. In some situations, regulations need to draw bright lines. A common example of a 'bright line' in hESCR is the rule that no research embryo may be allowed to develop *in vitro* beyond 14 days of development, or the appearance of the primitive streak. This commonly adopted, very specific restriction was put in place to address concerns about the moral status of embryos used for research. Although pro-life advocates argue that embryos have the full moral status of personhood, others believe that personhood is ascribed on the basis of certain characteristics. In the latter view, a collection of human cells that does not have sentience cannot be considered a person.[45] Before the development of the primitive streak, the embryo lacks the structure for neurological specialisation and hence cannot have sentience. For the purposes of this paper, the point is not to resolve the dilemma of the moral status of human embryos. Rather, that a general rule, for example a rule that stem cell researchers must respect the moral status of the embryo, is too vague to serve as a guide for institutional oversight committees or researchers.

Other research situations require a more complex balancing of flexibility and specificity. In the case of induced pluripotent stem cell (iPS cell) research, researchers derive pluripotent stem cells from somatic cells. Under US regulations, an exception to consent allows existing biological materials to be

---

[43] California Institute for Regenerative Medicine, *Regulations Title 17 California Code of Regulations*, Sections 100010–100110, 2006.

[44] Lomax GP, Hall ZH and Lo B. Responsible Oversight of Human Stem Cell Research: The California Institute for Regenerative Medicine's Medical and Ethical Standards, *Public Library of Science (PLoS) Medicine* **4**, 5 (2007): e114.

[45] National Institutes of Health, US. *Report of the Human Embryo Research Panel.* Bethesda, MD: National Institutes of Health, 1994.

used for research without consent if they are de-identified.[46] While neither the donation of materials to derive iPS cells nor their derivation raise special ethical issues, some potential downstream research uses of iPS cell derivatives may be highly sensitive, for example clinical allogeneic transplantation or reproductive research to create embryos. Some of the original somatic cell donors may not have agreed to such downstream research uses if their consent had been explicitly sought.[47] From a policy standpoint, there are conflicting ethical and policy considerations regarding such future uses of derivates of donated biological materials. On the one hand, researchers must respect donors and their wishes for how their biological material may be used in research. However, new iPS cell lines may be widely shared, and scientists deriving the lines cannot control additional research carried out with them. On the other hand, scientifically essential and ethically acceptable research should be encouraged in order to benefit society and future patients. Undue delays and administrative burdens on such research should be minimised. How can these countervailing considerations be reconciled? Researchers who derive iPS cell lines should use only somatic cells from donors who have given permission to carry out additional basic stem cell research, including genomic sequencing, injection into non-human animals, and sharing of lines with other researchers.[48] These types of non-clinical research are commonly carried out with donated biological materials without explicit consent from donors. However, because of the sensitivity of hSCR, specific consent is ethically desirable in this context. Moreover, it would be desirable to obtain permission to re-contact donors in the future to obtain consent for sensitive downstream research that could not be anticipated at the time of the original donation. This example of downstream stem cell research illustrates how flexible guidelines may allow novel ethical issues to be addressed through the development of specific professional or institutional standards rather than through highly prescriptive regulations.

RECOMMENDATIONS

In light of these differences in hESC policies around the world, what suggestions might be made regarding stem cell oversight in Singapore?

---

[46] Lo B. *Ethical Issues in Clinical Research: A Practical Guide*. Philadelphia, PA: Lippincot Williams & Wilkins, 2010, pp. 213–224.

[47] Aalto-Setälä K, Conklin BR and Lo B. Obtaining Consent for Future Research with Induced Pluripotent Cells: Opportunities and Challenges, *Public Library of Science (PLoS) Biology* **7**, 2 (2009): e1000042.

[48] *Ibid.*

Consistency with Other Countries

With increased globalisation, there is a need for harmonisation among stem cell policies in different countries. Ideally, lines derived in one country would meet requirements for use in other countries, for example regarding consent from donors whose materials were used to derive the lines. Such consistency would promote the sharing of stem cells among scientists in different nations. In turn, such consistency would advance the pace of scientific discoveries.

It would be useful to work towards the harmonisation of policies regarding stem cell research. A useful model is the international standards for good manufacturing practice and for good clinical practice, established through the International Conference on Harmonisation of Technical Requirements for Registration of Pharmaceuticals for Human Use, which facilitate simultaneous clinical trials and licensing of new drugs in multiple countries. These international standards are more detailed than, for example, the ethical guidelines for stem cell research adopted by the International Society for Stem Cell Research. Singapore, with its international orientation and its close scientific relationships with research institutions in many parts of the world, may be in a good position to advocate such harmonisation and to take a lead in its development.

More Explicit Policies

A light touch approach to regulation works best when all stakeholders, including those subject to regulation, share a common understanding of the main regulatory provisions and the principles that form the ethical justification for the regulations. Furthermore, it is useful if stakeholders share a common understanding of what specific actions and procedures are implied in the regulatory requirements. In some regulatory models, as in the US, there is much more specification of requirements, which can provide specific guides to decisions and actions compared to more general regulations, particularly for persons from other countries with different regulatory systems and cultural backgrounds.

With increasing globalisation, there will be scientists and companies from overseas carrying out research in Singapore or collaborating with research institutions in Singapore. It may not be feasible to assume that they understand the full implications of regulations that are framed in general language, as could be realistically assumed for scientists trained in Singapore or companies that were started in Singapore. At a minimum, it would be useful to have more detailed training for such expatriate investigators, preferably in an Internet-based format. Another option is to establish a certification process for researchers who are based outside of Singapore, to ensure that they understand the essential features of Singapore's stem cell regulations. Similar educational or certification requirements are in place in the US; for example, universities commonly require investigators to take an online course in research ethics and scientific misconduct and to pass a series of questions to demonstrate that they have learned the key points in the US federal regulations and the institutional policies.

Listening to Diverse Groups

As a multi-cultural society, Singapore has adopted a process of consultation with diverse religious groups on sensitive bioethics topics, including stem cell research. This model of outreach and consultation can foster trust in the regulatory process and acceptance of the final regulations. This process of outreach and consultation may also be useful with regard to countries with different cultural heritages and religious traditions. The goal of such discussions would not be necessarily to have all countries adopt the same regulations, but rather to persuade countries to defer to regulatory approval granted in other countries that have explicit stem cell policies that are consistent with universal ethical standards such as informed and voluntary consent and minimisation of risk to donors of materials used to derive stem cell lines, have provided public justification for those policies, and have carried out internal consultations with their public during the policy development process.[49] A practical goal would be to have countries accept review and approval of stem cell derivation carried out in other nations that meet these procedural standards, rather than trying to carry out *de novo* review for existing lines. Such acceptance of review in other countries would foster respect for reasonable differences of opinion among countries in interpreting and implementing universal ethical standards.

In summary, the rapidly developing field of hSCR requires effective oversight to ensure that it is carried out in an ethically appropriate manner. The US

[49] Lo B *et al.* Importing Human Pluripotent Stem Cell Lines Derived at Another Institution: Tailoring Review to Ethical Concerns, *Cell Stem Cell* **4** (2009): 115–123.

oversight of hSCR illustrates how oversight of this novel type of research can take advantage of existing frameworks of research oversight. As in other countries, the US oversight system is shaped by culture and attitudes towards government. At the same time, policies also are shaped by active choices among different models of regulation. Some policy makers addressing hSCR have tried to avoid problems that have emerged in other aspects of the research oversight system. In particular, the rapid scientific progress in hSCR underscores the need for flexibility in oversight. Some key policy makers and policy-making institutions have changed their views over time. Academic writings and best practices developed at research institutions have influenced national standards. Future policies regarding hSCR will benefit from continued attention to the advantages and disadvantages of different approaches to oversight.

ACKNOWLEDGEMENTS

The research for this chapter was supported by the US NIH Grant Number 1 UL1 RR024131 from the National Center for Research Resources (NCRR) and NIH Roadmap for Medical Research and by the Greenwall Foundation. Its contents are solely the responsibility of the authors and do not necessarily represent the official view of NCRR or NIH or the Greenwall Foundation. The authors have no conflicts of interest to declare. They thank Timothy Jost, JD, for helpful discussions regarding regulatory theories.

7

Genetics and Stem Cell Research: Models of International Policy-making

Bartha Maria Knoppers, Emily Kirby
and Rosario Isasi

INTRODUCTION: INTERNATIONAL NORMS

An examination of the historical background of international policy-making in the field of human genetics reveals that scientific advances often provoke public outcry, sometimes followed by reactive policy at both the international and the national levels. Lessons learned from the historical examination that follows underscore the need for scientific 'progress' to ideally be the subject of prospective policy deliberation.

In this chapter, we address the historical evolution of a number of international policy responses in the area of human genetics and stem cell research. The goal of this chapter is to provide a historical overview of the international policy landscape so as to contribute to current policy debates.

At the international level, our analysis is centred on the norms surrounding genetic research and therapy, cloning, and stem cell research developed by the Human Genome Organisation (HUGO), the World Health Organization (WHO), the Council for International Organizations of Medical Sciences (CIOMS), the United Nations Educational, Scientific and Cultural Organization (UNESCO), the United Nations (UN), and the Organisation for Economic Co-operation and Development (OECD). At the national level, we will use both Canada and Singapore as case studies on these topics. For historical accuracy, we have separated our international overview into the pre-1990 human genome project, then the period of 1990–2000, followed by the post-genomic era, before turning to Canada and Singapore, and a brief conclusion on current trends.

THE PRE-GENOMIC ERA: THE 1990S

International guidelines applicable to human genetics before the 1990s are few and none specifically address human genetic research. The main thrust of the 1990s was the elaboration of ethical norms translating the fundamental human rights of autonomy, privacy, and justice into the domain of biomedical research. The aim was to ensure the recognition of fundamental rights in the researcher-participant relationship. These rights combined with traditional, medical deontological principles "assimilat[ed] the researcher-participant relationship to the medical one albeit with a greater intensity of obligations to inform, to follow and to be monitored by ethics review boards".[1] Policy-making in biomedical ethics in this era followed the legacy of the Nuremberg Code (1948),[2] the Helsinki Declaration (1964),[3] and the Belmont Principles (1979).[4] It focused on enhancing the protection of human research participants and condemning research deemed unethical. Indeed, the *Proposed International Ethical Guidelines for Biomedical Research Involving Human Subjects* published by CIOMS (1982)[5] echoed the principles set forth in the Nuremberg Code. These guidelines were the first to propose the creation of 'ethical review committees', an innovative approach to reviewing research.[6] Thus, policy-making prior to 1990 was centred on protecting research participants from possible exploitation and harm.

---

[1] Knoppers BM. Genomics and Policymaking: From Static Models to Complex Systems? *Human Genetics* **125**, 4 (2009): 375–379, p. 375.

[2] *Trials of War Criminals before the Nuremberg Military Tribunals under Control Council Law No. 10*, Vol. 2, pp. 181–182. Washington, DC: US Government Printing Office, 1949.

[3] World Medical Association, *Declaration of Helsinki — Ethical Principles for Medical Research Involving Human Subjects*, 1964.

[4] National Commission for the Protection of Human Subjects of Biomedical and Behavioral Research, USA, *The Belmont Report: Ethical Principles and Guidelines for the Protection of Human Subjects of Research*, 1979.

[5] Council for International Organizations of Medical Sciences, *International Ethical Guidelines for Biomedical Research Involving Human Subjects*, 1982.

[6] Levine RJ. New International Ethical Guidelines for Research Involving Human Subjects. *Annals of Internal Medicine* **119**, 4 (1993): 339–341.

As the next sections also illustrate, another legacy of the pre-genomic period is the reliance on professional self-regulation by the medical research community. The proliferation of professional codes of conduct, consensus statements and declarations — as opposed to binding instruments in the following decades — demonstrate how the biomedical ethics model has continued to play a major role in the development of policy. That is, both clinicians and researchers took the initiative to develop professional norms rather than waiting for (or perhaps forestalling) external controls. These norms reveal the mechanisms for ensuring what they saw as a proper discharge of responsibility to patients or research participants. In terms of genetics research *per se*, the field of 'genethics' began to flourish following the start of the Human Genome Project in the 1990s.

1990–2000: THE HUMAN GENOME PROJECT

The beginning of the 1990s marks the instigation of the Human Genome Project, where policy-making trends follow both a stepwise and yet complex path, indicating a further conflation of the biomedical ethics model with the development of human rights generally around the world.[7] In particular, the CIOMS, WHO, HUGO, and UNESCO issued some of the early 'genethics' statements of this decade (directed at genetic research). This decade was also marked by an international debate on human reproductive cloning.

Genetic Research

As early as 1990, CIOMS adopted the *Declaration of Inuyama*[8] which offered a preview of the decade to come, exploring issues of growing concern about genetics and the human genome sequencing project, as well as genetic testing. The multidisciplinary team of experts drafting the *Declaration of Inuyama* recognised that the public perceived human genetics to be different from other medical information, hence the birth of

[7] Knoppers BM. Genomics and Policymaking: From Static Models to Complex Systems? *Human Genetics* **125**, 4 (2009): 375–379.
[8] Council for International Organizations of Medical Sciences. *The Declaration of Inuyama: Human Genome Mapping, Genetic Screening and Gene Therapy*, 1990.

genetic exceptionalism.[9] This understanding of human genetics greatly influenced policy-making and was expressed in the public perception that the knowledge generated from genetic research was the equivalent of exposing the personal 'code'. Indeed, "Public concern about the growth of genetic knowledge stems in part from the misconception that while the knowledge reveals an essential aspect of humanness it also diminishes human beings by reducing them to mere base pairs of deoxyribonucleic acid (DNA)".[10] This perception was expressed in national law and guidelines during this period.

The *Declaration of Inuyama* was one of the first statements to prospectively consider the socio-ethical and legal implications of genetics (for example, genetic testing, cloning, and gene therapy). Nevertheless, it openly rejected the above-mentioned genetic determinism.[11] Inspired by history, the *Declaration* recognised the deontological obligations of genetic researchers and therapists to ensure the ethical development and use of technologies. Thus, "By insisting on truly voluntary programmes designed to benefit directly those involved, they can ensure that no precedents are set for eugenic programmes or other misuse of the techniques by the State or private parties".[12] Notwithstanding the cautious approach of the *Declaration*, the drafters did not resort to premature, facile prohibitions, as exemplified by their stance towards issues such as germ cell therapy,[13] which were left for future consideration. Instead, the prospective and flexible approach of the *Declaration* called for continuous multidisciplinary and transcultural dialogue in the setting of ethical standards.

[9]Murray TH. Genetic Exceptionalism and "Future Diaries": Is Genetic Information Different From Other Medical Information? in *Genetic Secrets: Protecting Privacy and Confidentiality in the Genetic Era*, ed. M Rothstein. New Haven: Yale University Press, 1997, pp. 60–73.

[10]Council for International Organizations of Medical Sciences, *The Declaration of Inuyama: Human Genome Mapping, Genetic Screening and Gene Therapy*, 1990, Section II.

[11]Throughout the *Declaration*, the drafters reiterated that "efforts to map the human genome present no inherent ethical problems", and that "for the most part present genetic research and services do not raise unique or even novel issues, although their connection to private matters... and the rapidity of advances in genetic knowledge and technology, accentuate the need for ethical sensitivity in policy-making". See Council for International Organizations of Medical Sciences, *The Declaration of Inuyama: Human Genome Mapping, Genetic Screening and Gene Therapy*, 1990, Sections II and III.

[12]*Ibid*. Section VII.

[13]*Ibid*. Section VII.

Other CIOMS guidelines[14] of this decade, however, chose not to address genetics:

> *Certain areas of research do not receive special mention in these guidelines; they include human genetic research, embryo and fetal research, and fetal tissue research. These represent research areas in rapid evolution and in various respects controversial. The Steering Committee considered that since there is no universal agreement on all the ethical issues raised by these research areas it would be premature to try to cover them in the present guidelines.*[15]

However, with advances in the sequencing of the human genome, international organisations progressively came to understand the importance of research on genetic diseases and some statements began to address public health issues concerning inherited disease.[16] In this group of statements, the main goal, especially for the WHO, was to improve human health and prevent abuses:

> *The task of the Scientific Group was therefore: (1) to review the place of genetics in modern medicine; (2) to summarize current practical applications of genetic knowledge in the diagnosis, treatment and prevention of disease; (3) to consider the likely immediate impact of human genome research; (4) to help medical decision-makers to keep pace with these developments; and (5) to give guidance on the organization of genetic services.*[17]

In addition, this gene-disease focus maintained that "Advances in medical genetics should not be abused for non-medical purposes".[18]

---

[14]Council for International Organizations of Medical Sciences, *International Guidelines for Ethical Review of Epidemiological Studies,* 1991; and Council for International Organizations of Medical Sciences, *International Ethical Guidelines for Biomedical Research Involving Human Subjects,* 1991.

[15]Council for International Organizations of Medical Sciences, *International Ethical Guidelines for Biomedical Research Involving Human Subjects,* 1993, Background Note.

[16]World Health Organization, *Control of Hereditary Diseases — Report of a WHO Scientific Group,* 1996; World Health Organization, *Statement of the WHO Expert Advisory Group on Ethical Issues in Medical Genetics,* 1997.

[17]World Health Organization, *Control of Hereditary Diseases,* 1996, p. 2.

[18]*Ibid.* Point 14 of Conclusions and recommendations, p. 80.

There was also general recognition by the WHO of the importance of policy-making to address issues of ethical importance, including that of ensuring (in the long run) equitable access due to possible commercialisation and patenting: "The implications for genetic diagnosis and therapy of the patenting and ownership of DNA sequences, and of DNA technologies, are causing widespread concern; international agreement may be required to ensure that access to genetic technology, including gene therapy, is not restricted"[19]. International policy-making in this period continued to be characterised by the 'marriage' of human rights with biomedical ethics as normative instruments began to specifically address the context of genetic research.

For instance, during this decade, HUGO took the lead in the development of norms to protect research participants and to state basic principles of human rights in the context of genetics.[20] Interestingly, the principles articulated by HUGO were a preambular feature of all of its subsequent statements. These four principles were: "(1) recognition that the human genome is part of the common heritage of humanity, (2) adherence to norms of human rights, (3) respect for the values, tradition, culture, and integrity of participants, and (4) acceptance and upholding of human dignity and freedom".[21] Indeed, in its 1996 *Statement on the Principled Conduct of Genetics Research,* HUGO was the first to present the human genome "as part of the common heritage of humanity".[22] This *Statement* also established HUGO's recommendations for human genetic research[23] during this decade.

Likewise, UNESCO's 1997 *Universal Declaration on the Human Genome and Human Rights*[24] prospectively sets the framework for the protection of human dignity. It maintained that human dignity should prevail over any scientific endeavour, that "The human genome in its natural state shall not

---

[19] *Ibid.* Point 15 of Conclusions and recommendations, p. 80.

[20] Human Genome Organisation, *Statement on the Principled Conduct of Genetics Research,* 1996.

[21] *Ibid.*

[22] Wertz DC and Knoppers BM. The HUGO Ethics Committee: Six Innovative Statements, *GE*^{3}*LS* **2**, 1 (2003): 1. *GE*^{3}*LS* stands for genomics and its related ethical, economic, environmental, legal, and social aspects.

[23] The '10 C's', as listed by HUGO, are as follows: Competence, Communication, Consultation, Consent, Choices, Confidentiality, Collaboration, Conflict of interest, Compensation, Continual review. For more information, see: http://www.hugo-international.org/img/statment%20on%20the%20principled%20conduct%20of%20genetics%20research.pdf.

[24] United Nations Educational, Scientific and Cultural Organization, *Universal Declaration on the Human Genome and Human Rights,* 1997.

give rise to financial gain",[25] that human reproductive cloning should be prohibited,[26] and that as an act of international collaboration there should be a free exchange of scientific knowledge.[27] In that same year, the WHO proposed preliminary guidelines on the ethics of medical genetics.[28] This report served as a starting point for the discussion of general ethical issues of the application of genetic research (proper use of genetic testing and data gathering, genetic screening, patenting, etc.).

In 1998, HUGO's *Statement on DNA Sampling: Control and Access*[29] provided practical recommendations concerning consent and data protection (such as informed consent, coding, anonymisation, destruction mechanisms). It was one of the first to address issues with regard to the sampling, collection, and storage of DNA for genetic research. It cautioned against anonymisation of samples when detrimental to scientific objectives, even if this procedure would seemingly simplify legal and ethical obstacles to data and sample use.

Cloning

Turning to reproductive technologies during this same time frame, it should be noted that while the birth of Louise Brown in 1978 marked the success of *in vitro* fertilisation, it was not until 1997 with the cloning of Dolly the sheep that norms on human reproductive cloning emerged at the international level. Often *ad hoc* in nature, they included prohibitions and sometimes even the criminalisation of research activities such as reproductive cloning.

The first international prohibition on reproductive cloning was presented by the WHO in its 1998 *Ethical, Scientific and Social Implications of Cloning in Human Health.*[30] Interestingly, this statement was followed by a second on *Cloning in Human Health*[31] that examined not only cloning, but general gene manipulation techniques and their ethical implications for human health.

[25] *Ibid.* Article 4.
[26] *Ibid.* Article 11.
[27] *Ibid.* Article 19 (iv).
[28] World Health Organization, *Statement of WHO Expert Advisory Group on Ethical Issues in Medical Genetics*, 1997.
[29] Human Genome Organisation, *Statement on DNA Sampling: Control and Access*, 1998.
[30] World Health Organization, *Ethical, Scientific and Social Implications of Cloning in Human Health*, 1998.
[31] World Health Organization, *Cloning in Human Health: Report by the Secretariat (A52/12)*, 1999.

The scope of discussion was thereby extended to genetic manipulation while maintaining that reproductive cloning should be banned.

The 1999 HUGO *Statement on Cloning*[32] offered a flexible perspective by not altogether rejecting the use of cloning. In fact, while some forms of cloning were deemed unacceptable (eg. cloning for human reproductive purposes), other forms of cloning (eg. somatic cell therapeutic cloning) were considered to be interesting avenues for future therapies. Finally, this *Statement* also proposed that although creation of embryos solely for research was not acceptable, this situation could change in the case of a "widespread benefit for humanity".

In sum, during the 1990s, there seemed to be a consensus on banning certain reproductive cloning techniques but other techniques were not altogether proscribed and left open for future reflection. Overall, however, this period exhibited a linear or static approach to policy-making as demonstrated by the 'monogenic', one gene–one protein philosophy underlying the genetic determinist policies of this decade.[33] Increasing public familiarity with genomic research and mapping together with the scientific advances of the post-mapping era, however, shifted the trend towards a more dynamic approach.

2000 AND ONWARDS: THE POST-GENOMIC ERA

Genetic Research

Concomitant with the expansion of knowledge on the human genome following the completion of the sequencing project in 2001 (draft) and 2003 (complete), the number and content of normative policies increased. Issues addressed were related to the research use of genetic information and the increasing utilisation of databanks to store information. Therefore, norms no longer addressed the ethics of *permitting or prohibiting the use* of genetic data in medicine and health but, rather, issues such as the ethics of *how to use* and protect the security of such data.

[32] Human Genome Organisation, *Statement on Cloning*, 1999.

[33] Knoppers BM. Genomics and Policymaking: From Static Models to Complex Systems? *Human Genetics* **125**, 4 (2009): 375–379, p. 375.

Most policy statements of the post-genomic era reiterate the importance of genetics in research for public health and healthcare purposes. However, these statements acknowledge that the deciphering of the human genome goes beyond the monogenic model of genetic diseases that dominated the 1990s with its corresponding genetic exceptionalism.[34] A 2002 report by the WHO summarises this new approach: "It is now believed that the information generated by genomics will, in the long term, have major benefits for the prevention, diagnosis and management of many diseases which hitherto have been difficult or impossible to control."[35] Indeed, this was underscored in a subsequent 2005 WHO statement illustrating the evolving role of genetic research:

> *Increased knowledge of genomics over the past two decades has made it apparent that the traditional category of genetic diseases represents only those conditions in which the genetic contribution is particularly marked, whereas in fact diseases can be arrayed along a spectrum representing the varied contribution of genes and the environment. The beneficial applications of genomic knowledge are still evolving, but it is expected that in the future genomics will have "a significant contribution to make to the area of public health".*[36]

In addition, while maintaining an emphasis on developing consent and confidentiality and in providing transparency on possible future commercialisation, there was a continuing focus on the protection of research participants from potential harms.

Indeed, UNESCO's *International Declaration on Human Genetic Data* of 2003[37] proposed specific rules on consent, withdrawal of consent, and confidentiality, but again set genetic data apart from other medical information.

---

[34] Murray TH. Genetic Exceptionalism and "Future Diaries": Is Genetic Information Different From Other Medical Information? in *Genetic Secrets: Protecting Privacy and Confidentiality in the Genetic Era*, ed. M Rothstein. New Haven: Yale University Press, 1997, pp. 60–73.

[35] World Health Organization, *Genomics and World Health: Report of the Advisory Committee on Health Research*, 2002, p. 5.

[36] World Health Organization, *Control of Genetic Diseases: Report by the Secretariat (EB116/3)*, 2005, para 1.

[37] United Nations Educational, Scientific and Cultural Organization, *International Declaration on Human Genetic Data*, 2003.

In 2005, the *Universal Declaration on Bioethics and Human Rights*[38] continued this exceptionalist approach but also looked at societal issues such as benefit sharing and social responsibility. Particularly, it proposed a new perspective on the rules of ethics review, calling for "professionalism, honesty, integrity, and transparency in the decision-making process; the setting-up of ethics committees; appropriate assessment and management of risks; and ethical transnational practises that help in avoiding exploitation of countries that do not have an ethical infrastructure".[39] An interesting development in this 2005 *Declaration* was the elaboration of the concept of stigmatisation. The potential for labelling individuals or populations due to their genetic 'makeup', or due to the perception that these individuals at risk are already affected by their deleterious genes, prompted the adoption of the notion of stigmatisation and the implementation of measures aiming at its prevention.[40]

Another innovation of this era emanates from HUGO's *Statement on Benefit Sharing.*[41] This was the first policy to apply a benefit-sharing model to the human genome *per se* categorising it as a "shared resource" for the benefit of all humanity. These issues were reiterated in HUGO's *Statement on Human Genomic Databases,*[42] which considered primary genomic sequences as "global public goods". Again, its 2007 *Statement on Pharmacogenomics: Solidarity, Equity and Governance*[43] called for international and national governance for the sharing and protection of data concerning pharmacogenomics as well as the need to integrate knowledge from research into clinical practice. Similarly, WHO's *Genetic Databases: Assessing the Benefits and the Impact on Human and Patient Rights*[44] also addressed the issue of data sharing from a human rights perspective, the objective being to promote public health.

---

[38] United Nations Educational, Scientific and Cultural Organization, *Universal Declaration on Bioethics and Human Rights,* 2005.

[39] *Ibid.* Article 18.1.

[40] Rivard G. Article 11: Non-Discrimination and Non-Stigmatization, in *The UNESCO Universal Declaration on Bioethics and Human Rights: Backgrounds, Principles and Applications,* eds. HAMJ ten Have and MS Jean. Paris: UNESCO Publishing, 2009, pp. 187–198.

[41] Human Genome Organisation, *Statement on Benefit Sharing,* 2000.

[42] Human Genome Organisation, *Statement on Human Genomic Databases,* 2002.

[43] Human Genome Organisation, *Statement on Pharmacogenomics: Solidarity, Equity and Governance,* 2007.

[44] World Health Organization, *Genetic Databases: Assessing the Benefits and the Impact on Human and Patient Rights,* 2003.

The patenting of DNA, as discussed in the WHO's 2005 report on *Genetics, Genomics and the Patenting of DNA: Review of Potential Implications for Health in Developing Countries,*[45] was a related issue of concern. In that report, expressions such as "genetics as the common heritage of mankind" as well as genes as "public goods" were revisited in terms of commercialisation and patenting, and the impact of the various past policies are re-examined.[46] The WHO concluded that patenting DNA could restrict access to genetic information and hinder collaboration in research and clinical applications.

The CIOMS updated its 1882 guidelines in 2002,[47] with a further update for epidemiological research in 2009.[48] The 2002 guidelines were not specific to genetic research; rather, general aspects of the protection of research participants were developed. The 2009 update shifted the scope of the guidelines from a 'biomedical research' focus to an approach centred on 'epidemiological studies', including large scale population studies. This was in all likelihood due to the emergence of such studies around the world.[49] In terms of genetic research, the 2009 guidelines predominantly address issues relating to confidentiality and disclosure of results from genetic tests to relatives and third parties.

Finally, being also concerned about the protection of human subjects, the OECD developed in 2007 a document pertaining to molecular genetic testing.[50] These guidelines propose international standards, directed mainly at national governing bodies, on procedures to be used in the area of clinical genetic testing for the diagnosis of conditions and diseases. However, the text also refers to the various UNESCO *Declarations* on genetics with regards to patients' rights and protection. This was reinforced in its 2009 *Guidelines for Human Biobanks and Genetic Research Databases*[51]

---

[45]World Health Organization, *Genetics, Genomics and the Patenting of DNA: Review of Potential Implications for Health in Developing Countries,* 2005.
[46]*Ibid.*
[47]Council for International Organizations of Medical Sciences, *International Ethical Guidelines for Biomedical Research Involving Human Subjects,* 2002.
[48]Council for International Organizations of Medical Sciences, *International Ethical Guidelines for Epidemiological Studies,* 2009.
[49]See for instance the Public Population Project in Genomics (P3G). Information available at: http://www.p3g.org/secretariat/index.shtml.
[50]Organization for Economic Co-operation and Development, *OECD Guidelines for Quality Assurance in Molecular Genetic Testing,* 2007.
[51]Organization for Economic Co-operation and Development, *OECD Guidelines for Human Biobanks and Genetic Research Databases,* 2009.

in which governance mechanisms as well as legal and ethical guiding principles for population-based biobanks are proposed. Recognising that knowledge in the evolving domain of human genetics will depend on bringing together different strands of information through the use of databases, the OECD tailors its principles to the use of information for research purposes.

Cloning, Gene Therapy and Stem Cell Research

Whereas the main area of 'generalised ethical concern' during the 1990s was centred on emerging cloning techniques, the next decade was marked by a continued fear of its implications, together with guidance on gene therapy and stem cell research.[52]

First, there was a re-examination of the cloning debate in light of the development of new techniques in the field of 'research cloning'. In the early 2000s, an increased awareness of the shortcomings of legislation on reproductive cloning[53] led to a quasi-international unanimity on the dangers of reproductive cloning,[54] though somatic cell nuclear transfer (SCNT) had yet to receive significant international consensus. In order to answer this pressing need, the UN drafted its *Declaration on Human Cloning*[55] after several years of negotiations. The *Declaration* "... [was] drafted in a way to reflect the broad social concerns against species-altering technologies and to close loopholes in existing legal documents and declarations... The time [was] ripe for a flat-out international ban on human cloning".[56] The 2005 UN *Declaration* had a broader, international scope with consensus on the prohibition of

---

[52]See Human Genome Organisation, *Statement on Gene Therapy and Research*, 2001; and Human Genome Organisation, *Statement on Stem Cells*, 2004.

[53]Annas GJ, Andrews LB and Isasi RM. Protecting the Endangered Human: Toward an International Treaty Prohibiting Cloning and Inheritable Alterations, *American Journal of Law & Medicine* **28**, 2&3 (2002): 151–178.

[54]Isasi RM and Annas GJ. Arbitrage, Bioethics and Cloning: The ABCs of Gestating a United Nations Cloning Convention, *Case Western Reserve Journal of International Law* **35** (2003): 397–414.

[55]United Nations, *Declaration on Human Cloning*, GA Res. 59/280, UN GAOR, 59th Sess., UN Doc. A/RES/59/280, 2005.

[56]Annas G, Andrews LB and Isasi RM. Protecting the Endangered Human: Toward an International Treaty Prohibiting Cloning and Inheritable Alterations, *American Journal of Law & Medicine* **28**, 2&3 (2002): 151–178, p. 172.

reproductive cloning but not on research cloning. In any event, a *Declaration* is not legally binding though it can be extremely influential (eg. the *Universal Declaration of Human Rights*[57]).

Following a United Nations University — Institute of Advanced Studies report on the governance of human reproductive cloning in 2007,[58] UNESCO's International Bioethics Committee (IBC) was mandated to reexamine the ban on reproductive cloning in light of new scientific developments and possible change of perspectives on the issue. The IBC's 2009 *Report on Human Cloning and International Governance*[59] calls for the creation of a clear, legally binding instrument since the UN's 2005 *Declaration on Human Cloning*[60] was deemed insufficient to prevent human reproductive cloning. However, even in 2009, UNESCO was still unable to move forward with the adoption of an enforceable normative instrument and instead called for "dialogue on the international governance of human cloning" as "essential to foster public sensitivity and awareness-raising, with special attention to developing countries".[61]

A second area of international discussion was gene therapy. Already in 2001, HUGO's *Statement on Gene Therapy Research* stated that there was an important shift in the purposes of this technique:

> *Some 15 years ago, it was generally believed that the chief focus of gene therapy would be single gene diseases. Gene therapies for immune deficiencies, inherited anaemias and cystic fibrosis, for example, are the subjects of active research. Emphasis has shifted to attempts at experimental gene therapy for eventual use for common multigenic disorders, such as cancers and cardiovascular disease.*[62]

[57] United Nations, *Universal Declaration of Human Rights*, GA Res. 217 (III), UN GAOR, 3rd Sess., Supp. No. 13, UN Doc. A/810 at 71, 1948.
[58] United Nations University — Institute of Advanced Studies, *Is Human Reproductive Cloning Inevitable: Future Options for UN Governance*, 2007.
[59] United Nations Educational, Scientific and Cultural Organization — International Bioethics Committee, *Report of IBC on Human Cloning and International Governance*, June 2009.
[60] United Nations, *Declaration on Human Cloning*, GA Res. 59/280, UN GAOR, 59th Sess., UN Doc. A/RES/59/280, 2005.
[61] United Nations Educational, Scientific and Cultural Organization, *Conclusions of the 16th Session of the Intergovernmental Bioethics Committee*, 10 July 2009.
[62] Human Genome Organisation, *Statement on Gene Therapy and Research*, April 2001.

Furthermore, in contrast to other international organisations, except CIOMS, HUGO did not close the door to discussions concerning germline research. Rather, it proposed future attempts to examine different types of gene therapy research while incorporating public concerns about their potential.

The debates concerning cloning, gene therapy, and stem cell research are not unrelated. HUGO's *Statement on Stem Cells*[63] encourages advances in this area. For instance, it states that research concerning stem cells should not focus exclusively on spare embryos but should include different types of 'created' embryos (for example, somatic cell nuclear transfer 'embryos'). Without providing any explicit rationale, neither CIOMS nor the WHO has addressed stem cell research.

Two international bodies however have provided guidance in this area. The first is the International Stem Cell Forum (ISCF).[64] The overall aim of the ISCF is to promote global good practices through collaboration in stem cell research. Its Ethics Working Party (EWP) is mandated to identify, anticipate, and analyse ethical and policy issues involved in stem cell research and to develop and maintain an overview of these issues at the national and international level.[65] Furthermore, the mission of the EWP is to facilitate international dialogue on ethical issues and to foster the identification of shared ethical principles so as to guide the conduct of stem cell research and policy-making. To this end, the EWP has produced consensus opinions on payment for procurement of oocytes,[66] procedural safeguards for stem cell research,[67] and a rationale for a registry of clinical stem cell trials.[68]

The International Society for Stem Cell Research (ISSCR) is another initiative to foster global governance and facilitate international cooperation in

[63] Human Genome Organisation, *Statement on Stem Cells*, 2004.
[64] For more information, see: www.stemcellforum.org.
[65] *Ibid.*
[66] Isasi RM and Knoppers BM. Monetary Payments for the Procurement of Oocytes for Stem Cell Research: In Search of Ethical and Political Consistency, *Stem Cell Research* **1** (2007): 37–44; and Knoppers BM *et al.* (For the ISCF Ethics Working Party). Letter to the Editor: Oocyte Donation for Stem Cell Research, *Science* **316** (2007): 368–370.
[67] Knoppers BM *et al.* (For the ISCF Ethics Working Party). Ethics Issues in Stem Cell Research, *Science* **312** (2006): 366–367.
[68] Isasi RM and Nguyen MT. The Rationale for a Registry of Clinical Trials Involving Human Stem Cell Therapies, *Health Law Review* **16**, 2 (2008): 56–68.

stem cell research and banking. ISSCR seeks the harmonisation of core ethical principles through the adoption of professional guidelines. The guidelines seek to promote responsible, transparent, and uniform practices worldwide. The ISSCR's 2006 *Guidelines for the Conduct of Human Embryonic Stem Cell Research*[69] address issues of procurement, banking, derivation, distribution, and use of stem cells. Furthermore, in 2008, the ISSCR moved its focus from basic research to clinical translation, by promulgating the first professional guidelines setting standards for the clinical translation of stem cells[70] in three major areas: cell processing and manufacturing, pre-clinical studies, and clinical research. The guidelines encompass recommendations and establish principles for the conduct of translational stem cell research. Overall, ISSCR's guidelines adopted a classic biomedical ethics model by focusing heavily on informed consent, privacy, justice, and ethics review committees. The ISSCR strongly encourages the enactment of national laws that implement these guidelines. In the absence of such a legislative framework, the ISSCR proposes implementation through a mechanism of ethics review, which entails statements of compliance to be developed by intermediate agencies and actors (such as journal editors and granting agencies).

The preceding historical account of international policy making illustrates how scientific advances in the area of human genetics have inspired and generated a continual refinement of the biomedical ethical principles governing research. The reflections below on the Canadian policy landscape, followed by Singapore, provide an interesting example of how these historical, political, and scientific norms shape national initiatives.

Canada

In Canada, guidelines concerning genetics and stem cell research have existed since 1987 and were adopted by the Medical Research Council of Canada (MRC).[71] This Council provided the initial leadership for the later

[69] International Society for Stem Cell Research, *Guidelines for the Conduct of Human Embryonic Stem Cell Research*, 2006.

[70] International Society for Stem Cell Research, *Guidelines for the Clinical Translation of Stem Cells*, 2008.

[71] Medical Research Council of Canada, *Guidelines on Research Involving Human Subjects*, 1987.

drafting of the *Tri-Council Policy Statement* (TCPS)[72] adopted in 1998. In 2000, the Canadian Institutes of Health Research (CIHR) created its own ethics office and took over the mandate of the MRC. The TCPS proposes trans-disciplinary guidelines applicable to all forms of research involving humans, including research in the social sciences and the humanities. The guidelines set ethical norms to delimit the duties and rights of all those implicated in research involving humans. A section of the TCPS addresses human genetics specifically[73] and another delimits acceptable research regarding human gametes, embryos and foetuses.[74] For instance, gene alterations, the creation of human embryos for the purpose of research as well as reproductive cloning are prohibited.[75] However, the TCPS also provides additional norms on consent, privacy, return of results, and genetic counselling regarding human genetic research.[76] Currently, the TCPS is the main national reference for all publicly funded bodies undertaking research involving humans.

In 2008[77] and then again in 2009[78] a draft — the first attempt at a comprehensive revision since its 1998 adoption — became available for public comment. A new version of the TCPS is expected in 2010. The 2009 draft version emphasises respect for human dignity and the proportionality of ethics review. Furthermore, the original principles of the 1998 version have been distilled into three core principles: concern for welfare, respect for autonomy, and respect for the equal moral status of all humans.[79] In addition, the guidelines deal with specific issues of privacy, confidentiality, secondary uses of research data or materials, and stem cell and human genetic research.

[72] Canadian Institutes of Health Research, Natural Sciences and Engineering Research Council of Canada, Social Sciences and Humanities Research Council of Canada, *Tri-Council Policy Statement: Ethical Conduct for Research Involving Humans*, 1999 (with 2000, 2002, 2005 amendments).
[73] *Ibid.* Sections 8.1–8.8.
[74] *Ibid.* Sections 9.1–9.4.
[75] *Ibid.* Section 9.4.
[76] *Ibid.* Sections 8.2 and 8.3.
[77] Interagency Advisory Panel on Research Ethics, Canada, *Draft 2nd Edition of the Tri-Council Policy Statement: Ethical Conduct for Research Involving Humans*, 2008.
[78] Interagency Advisory Panel on Research Ethics, Canada, *Revised Draft 2nd Edition of the Tri-Council Policy Statement: Ethical Conduct for Research Involving Humans*, 2009.
[79] See Interagency Advisory Panel on Research Ethics, Canada, *What's New in the TCPS?* Available at: http://pre.ethics.gc.ca/policy-politique/initiatives/docs/What's%20New%20in%20the%20TCPS.pdf.

Based on the general principles set out in the 1998 TCPS, the CIHR developed the *Best Practices for Protecting Privacy in Health Research*[80] in 2005. It addressed certain issues concerning the dissemination of genetic information:[81]

> *Researchers, particularly those in the areas of health services, population and public health, and genetics/genomic research who study whole populations, should strive to communicate with the relevant population and governmental authorities regarding results that are pertinent to the improvement of health and/or the prevention of disease. The population studied should be made aware of possible socio-economic discrimination or group stigmatization as a result of the research results, such as because of perceptions of genetic risks. In the context of genetic research, the population should also be informed of the means taken to minimize the risks.*

The CIHR also addressed stem cell research. Its *Guidelines for Human Pluripotent Stem Cell Research*[82] were first drafted in 2002, with updates in 2006 and 2007. This set of *Guidelines* was put into place to further interpret and expand the TCPS guidelines. Their overall objective is to ensure the ethical and scientific oversight of human pluripotent stem cell research by setting the parameters of its permissibility (eg. use of embryos in stem cell research and creation of human embryos[83]). As with the TCPS, the CIHR *Guidelines* are based on several guiding principles: potential benefit for Canadians, respect for autonomy (ie. informed consent), privacy, and confidentiality. These core principles also contain a ban on commodification (eg. prohibition of commercialisation of human reproductive materials and on financial incentives for donation) and instrumentalisation (eg. creation of embryos for research purposes). They further call for "respect for individual and community notions of human dignity and physical, spiritual and cultural integrity".[84]

---

[80] Canadian Institutes for Health Research, *CIHR Best Practices for Protecting Privacy in Health Research*, September 2005.
[81] *Ibid.* p. 9 (emphasis in original).
[82] Canadian Institutes for Health Research, *Updated Guidelines for Human Pluripotent Stem Cell Research*, 27 June 2007.
[83] *Ibid.* Sections 8.1, 8.2.
[84] *Ibid.* Section 4.0.

In addition, these *Guidelines* cover other issues such as consent, confidentiality, and commercial interests in this area. They support stem cell research in general as it "holds great potential to treat human disease",[85] but limits their derivation to an acceptable source (ie. surplus embryos). Most importantly, the *Guidelines* exceptionalise such research through the creation of a special ethics review board — the Stem Cell Oversight Committee (SCOC) — charged to approve all research proposals dealing with human pluripotent stem cell research.[86]

Finally, following a Royal Commission report in 1993 and a further decade of discussion, a law to address assisted human reproduction and genetic research came into force in 2004. It is currently the only nationally applicable and legally binding instrument in Canada. The *Act Respecting Assisted Human Reproduction and Related Research (AHR Act)*[87] addresses issues such as informed consent, cloning, and the use of genetics and *in vitro* technologies to assist human reproduction. It sets out both prohibitions and regulatory requirements that are backed by criminal sanctions. For instance, some practices regarding reproductive and research cloning, altering the genome of a human or embryo, creating chimeras, paying an individual to act as a surrogate mother, and purchasing gametes in Canada, are considered criminal activities. Sanctions include severe fines and even imprisonment. Other practices are subject to regulation and include genetic research, prenatal and preimplantation diagnoses, donor registration, and accreditation of clinics under a new federal regulatory agency called Assisted Human Reproduction Canada. While not contesting the criminal activities prohibited under the AHR Act, the province of Quebec considers the other regulatory domains of the AHR Act to be unconstitutional since 'health' is a provincial matter. A reference case is pending before the Supreme Court of Canada.[88] Until an

---

[85] *Ibid.* Section 1.0.

[86] The scope of application of the CIHR guidelines encompasses new or ongoing human stem cell research that is:

1. funded by three federal granting agencies (CIHR, the Natural Sciences and Engineering Research Council, and the Social Sciences and Humanities Research Council);
2. conducted under the auspices of an institution that receives any Agency funding, whether on site or off site; or
3. conducted elsewhere with any source of funding, by faculty, staff, or students from an institution that receives Agency funding.

[87] Canada: *Act Respecting Assisted Human Reproduction and Related Research*, c.2, S.C. 2004.

[88] Renvoi fait par le gouvernement du Québec en vertu de la Loi sur les renvois à la Cour d'appel, L.R.Q., ch. R-23, relativement à la constitutionnalité des articles 8 à 19, 40 à 53, 60, 61 et 68

opinion is provided by the Supreme Court of Canada on the constitutionality of parts of the AHR Act, Health Canada — the agency responsible for developing regulations pursuant to the legislation, has decided to delay the prepublication of draft regulations. Consequently, there is no direction as per how important provisions of the AHR Act should be interpreted so a great degree of policy uncertainty remains.

For instance, under the terms of both the AHR Act and the CIHR Guidelines, no payments can be made to sperm or egg donors for their donation, except for the reimbursement of reasonable out-of-pocket receipted expenditures; and this, only if provided by a clinic with a license and according to regulations. As the regulations are yet to be developed, there is no compensation being provided to egg and sperm donors, and consequently no donors have come forward.[89] Moreover, paradoxically, gametes are imported and paid for from the USA with a resulting increase in commercialisation in the reproductive arena! Indeed, contrary to the proactive approach taken by Singaporean policymakers, in Canada there has not been a proactive analysis of the implications of the current regulatory system in the context of stem cell research.[90]

In sum, akin to Singapore's approach to genetic and stem ceil research policy which is the subject of the next section, the Canadian framework is a combination of public and private ordering. However, the policy models adopted by these jurisdictions show significant divergence. Indeed, Canada has adopted a more restrictive approach to the regulation of research. While a number of technologies and research areas are allowed and closely controlled (eg. gene therapy, genetic research, stem cells) by modest state intervention, others have been not only prohibited but are considered a serious criminal offence (eg. cloning). Influenced by fears concerning potential exploitation of women, commodification of the human body, and slippery slopes (ie. research on reproductive cloning), Canada has adopted an intermediate

---

de la Loi sur la procréation assistée, L.C. 2004, ch. 2 (Dans l'affaire du), 2008 QCCA 1167 (CanLII) (appealed, Suprême Court of Canada, *Attorney General of Canada* v. *Attorney General of Quebec*, 2008-08-26, 32750).

[89] Canadian Medical Association. Sperm Donor Pool Shrivels When Payments Cease, *Canadian Medical Health Journal* **182**, 3 (2010): 233.

[90] Both the scope of the public consultations and the draft regulations pertaining to the AHR Act have focused only on assisted reproductive services, leaving aside any reference to the research field. See, for instance, Health Canada, *Reimbursement of Expenditures under the Assisted Human Reproduction Act*, Public Consultation Document.

approach to policy-making. The result of political compromises and trade-offs that sought to balance diverse interests and values — if not to thwart the adoption of any legal framework — the Canadian model has created a system that is often internally inconsistent and unable to adapt to the rapid pace of scientific discoveries.[91]

Moreover, the influence and the implementation of international policies into Canadian practices have had some shortcomings. For instance, Canadian norms concerning genetic research seem to focus mainly on 'individualistic ethics' (respect for autonomy and privacy),[92] whereas international policies are going beyond this point and developing policies promoting collaboration and the free flow of data amongst the research community especially in population studies. In Canada, "[Research Ethics Boards] are uncomfortable with the "common goods" and public health nature of such international collaboration population infrastructures".[93] Therefore, conforming Canadian practices to international guidelines has not been straightforward. For instance, "while the OECD is attempting to provide preliminary guidance for countries involved in such efforts, Canadian REBs are less prepared for the implications of ethics of solidarity and reciprocity underlying the building of these research tools where international access and use are the norm".[94] Perhaps the ensuing consultation process will begin to address concerns about the implementation of international ethics norms. Indeed, the 2009 draft of the TCPS states in its introduction:[95]

> *This Policy expresses the Agencies' continuing commitment to the people of Canada to promote the ethical conduct of research involving human participants. It has been informed, in part, by leading international ethics norms, all of which may help, in some*

---

[91] See, for instance, Rugg-Gunn PJ *et al.* The Challenge of Regulating Rapidly Changing Science: Stem Cell Legislation in Canada, *Cell Stem Cell* **3**, 4 (2009): 285. See also Bordet S *et al.* The Changing Landscape of Human-Animal Chimera Research: A Canadian Regulatory Perspective, *Stem Cell Research* **4** (2010): 10.

[92] See, for example, Canadian College of Medical Geneticists, *Guidelines for DNA Banking*, 2008.

[93] Knoppers BM. Challenges to Ethics Review in Health Research, *Health Law Review* **17**, 2&3 (2009): 47.

[94] *Ibid.*

[95] Interagency Advisory Panel on Research Ethics, Canada, *Revised Draft 2nd Edition of the Tri-Council Policy Statement: Ethical Conduct for Research Involving Humans*, 2009, Introduction, p. 1.

measure, to guide the conduct of human research in Canada and by Canadian researchers abroad.

Overall, then, there is a strong tendency to integrate various disciplines into the ethical debate (as demonstrated, for example, by the breadth of the TCPS)[96] in Canadian policy. But there is a multitude of oversight committees and bodies at different levels (eg. local ethics boards and regional committees) which creates problems in the consistent application of ethical norms. Indeed, "The 'dysregulation' of human subjects research with inflexible requirements for adherence to narrow interpretations of every word in regulations and "guidance" policies has created a stranglehold and a loss of respect for ethics by scientists".[97] As just illustrated, the response to these complaints concerning the Canadian research ethics boards has been to review their role and functioning in the recent draft versions of the TCPS.

As we now turn to Singapore, it is interesting to note that its policy recommendations explicitly refer to international norms as well as to those of other countries (eg. Canada). Perhaps this is an indication of international and transnational cross-fertilisation.

Singapore

In 2000, Singapore created the Bioethics Advisory Committee (BAC) to examine the ethical, legal, and social issues arising from biomedical research and to recommend policies. This is not to say that there was an absence of norms in the preceding decade. Indeed, the *Medicines Act* of 1975 (Medicines Act),[98] regulations on clinical trials (Clinical Trials Regulations),[99] and the *Singapore Guideline for Good Clinical Practice* (SGGCP)[100] together provided a regulatory framework for pharmaceutical drug trials.

In addition, respect for the *Declaration of Helsinki*[101] already mandated ethics review of biomedical research involving humans and was ensured by

[96] Knoppers BM. Challenges to Ethics Review in Health Research, *Health Law Review* **17**, 2&3 (2009): 47.
[97] *Ibid.*
[98] Singapore Statutes: *Medicines Act* (Cap. 176), Revised 1985.
[99] Ministry of Health, Singapore, *Medicines (Clinical Trials) Regulations*, 27 March 1978.
[100] Ministry of Health, Singapore, *Singapore Guideline for Good Clinical Practice*, 1999 (revised).
[101] World Medical Association, *Declaration of Helsinki — Ethical Principles for Medical Research Involving Human Subjects*, 1964.

the guidelines of the National Medical Ethics Committee (NMEC), including its *Ethical Guidelines on Research Involving Human Subjects* of 1997 (NMEC Ethical Guidelines),[102] as well as those specifically on gene technology in 2001 (Gene Technology Guidelines).[103] In addition, the BAC has issued recommendations on reproductive and therapeutic cloning in 2002.[104] Finally, in 2004, the BAC's report on *Research Involving Human Subjects: Guidelines for IRBs* (IRB Guidelines)[105] followed prior recommendations raised in its consultation paper of 2003[106] and provided further direction. More specifically, it outlined the types of biomedical research requiring independent ethics review as well as the different options regarding this review (full, expedited, or exempted). The IRB Guidelines were also proposed for the creation, role, and duties of IRBs, and the respective responsibilities of research and institutions towards these committees were examined. These IRB Guidelines are consistent with the international ethical norms governing biomedical research. Indeed, it should be mentioned that what is unique about the work of the BAC is that, contrary to the work of most national bodies, there is explicit and often extensive reference to both international instruments and those of other countries. It could be said that this reflects an openness and willingness to learn from and build on acquired expertise from other ethics advisory bodies. In this sense, these policy documents reflect a level of consensus that is enriched not only by prior local consultation with the population of Singapore but also by socio-ethical and cultural international normative diversity.

Genetic research and gene therapy

In Singapore, there is no legislation explicitly regulating gene therapy; however, as mentioned in the introduction to this section, there is a regulatory

[102] National Medical Ethics Committee, Singapore, *Ethical Guidelines on Research Involving Human Subjects*, August 1997.
[103] National Medical Ethics Committee, Singapore, *Ethical Guidelines for Gene Technology*, February 2001.
[104] Bioethics Advisory Committee, Singapore, *Ethical, Legal and Social Issues in Human Stem Cell Research, Reproductive and Therapeutic Cloning*, June 2002.
[105] Bioethics Advisory Committee, Singapore, *Research Involving Human Subjects: Guidelines for IRBs*, November 2004.
[106] Bioethics Advisory Committee, Singapore, *Advancing the Framework on Ethics Governance for Human Research: A Consultation Paper*, 16 September 2003.

system in place which provides ethical governance of clinical trials. Concerning drug trials, the *Medicines Act*, the *Clinical Trials Regulations*, and the *SGGCP* provide the current formal regulatory framework.

For clinical research other than drug trials, three main committees have been created to provide bioethical infrastructure for human research in Singapore. In 1994, the MOH established the NMEC, a national policy advisory board which provides guidance regarding ethical issues in medical practice and establishes high standards of ethical research in Singapore. The NMEC has issued several guidelines which include the NMEC *Ethical Guidelines* and its *Gene Technology Guidelines*. In 1998, the MOH announced the acceptance of the NMEC *Ethical Guidelines* and required that all IRBs from public or restructured hospitals comply with the established guidelines.

Mirroring international policy development during the 'post-genomic era', the NMEC adopted comprehensive recommendations[107] regarding genetic testing, gene transfer, somatic and germ-line gene therapy in the context of both clinical and research practice. As with other policy statements developed in this era, the NMEC recommendations highlight the importance of genetics in research and medical practice while enhancing protections for both research participants and patients.

In this context, the NMEC's *Gene Technology Guidelines* established a framework of ethical principles for 'gene technology', which has been defined as "the use of techniques for the analysis and/or manipulation of DNA (deoxyribonucleic acid), RNA (ribonucleic acid) and/or chromosomes".[108] Regarding genetic testing (ie. clinical genetic tests, newborn screening, antenatal and carrier testing), this framework included comprehensive and prospective recommendations for protecting the autonomy of both patients and research subjects (eg. provisions for genetic counselling, informed consent) as well as for their confidentiality and privacy. It further provided measures for preventing genetic discrimination in insurance and employment and for the regulation of the advertisement and marketing of genetic technologies.

As concerns gene therapy (gene transfer, somatic and germ-line therapy), the *Gene Technology Guidelines* established a governance framework for

[107] National Medical Ethics Committee, Singapore, *Ethical Guidelines for Gene Technology*, February 2001.
[108] *Ibid.* Section 1.1.

research and clinical applications. They highlighted important scientific distinctions between gene therapy and gene transfer, as is exemplified by the following statement: "... there are important applications of gene transfer in clinical research which are not truly therapeutic ... It is important from the ethical standpoint to appreciate that subjects in such gene transfer experiments do not directly benefit from these procedures".[109] The NMEC supported somatic gene therapy but recommended that germ-line gene therapy, which transmits genetic changes to offspring in subsequent generations, should not be permitted. The NMEC further stated that somatic gene therapy is fundamentally not different from any form of organ transplantation, blood cells transfusion, or experimental treatment.

The ethical and governance framework proposed by the NMEC included mandating peer review of research protocols on gene technology applications and a rigorous evaluation of the benefits and risks involved before their clinical application. As with their recommendations for genetic testing, the NMEC *Guidelines* established ethical safeguards to protect patients and research subjects (eg. recommendations for vulnerable populations, mandatory follow-up) and for enhancing the informed consent process. Moreover, following international practice, the Committee recommended that "gene therapy in humans should be confined to alleviating disease in individual patients"[110] and that gene therapy to enhance or change normal traits should be strictly prohibited.

Following this prospective approach to policy-making, the BAC issued a report on *Personal Information in Biomedical Research* in 2007 (Personal Information Report).[111] While not specifically on genetic research, it is this very fact that sets Singapore apart. Personal information is very broadly defined as "any information about an individual ... includes personal particulars, details of medical conditions and healthcare management, physical or psychological measures, dietary requirements and religious or other beliefs".[112] While cognisant that ethical difficulties arise when predictive information is involved, genetic information is not distinguished from other medical information (ie. genetic exceptionalism) but rather, considering public unease and the still uncertain actuarial precision and value of

[109] *Ibid.* Section 5.3.
[110] *Ibid.* Summary of recommendations, para 21.
[111] Bioethics Advisory Committee, Singapore, *Personal Information in Biomedical Research*, May 2007.
[112] *Ibid.* para 2.1, p. 12.

such information, a moratorium on the use of predictive genetic tests results by insurers is simply recommended. This report is also worth highlighting for several reasons — both principled and practical. As concerns principles, the concept of a proportional approach to the issue of general versus specific consent to research and in the balancing of privacy and confidentiality and, secondly, that of reciprocity in the use by health authorities of personal information are put forward. The application of both these principles has important implications for genetic research.

First, information that is irreversibly de-identified is not subject to privacy and confidentiality requirements. Second, general consent is sufficient for subsequent unspecified research, if samples and data are coded. Third, the authorities should

> *clarify the legal basis for the disclosure of medical information to national disease registries by physicians and establish mechanisms enabling national registries and healthcare institutions to facilitate the use of personal information ... for biomedical research that can significantly advance the public good, while safeguarding privacy.*[113]

Furthermore, IRBs should be legally empowered "to waive the patient consent requirement for research involving only the use of medical records, while ensuring patient privacy and confidentiality of medical information".[114] The Personal Information Report recommends a statutory framework for the protection of personal information. This Report puts Singapore at the forefront of ethical guidance specific to research involving data and consent and privacy issues as distinct from the context of clinical trials. The same 'avant-garde' thinking is evident in its policy positions on cloning and stem cell research.

Cloning and stem cell research

In Singapore, as in many jurisdictions across the globe, the debate on human reproductive cloning has been influential in the framing of policy responses. Indeed, both the first consultation paper and report issued by the BAC centred on human stem cell research and cloning. Akin to the Canadian

---

[113] *Ibid.* Recommendation 7, pp. 7 and 33.
[114] *Ibid.* Recommendation 8, pp. 8 and 34.

policy process, policy-making activities in Singapore took place after lengthy public consultation processes[115] that aimed at engaging a wide range of stakeholders (eg. religious, patient, professional, research, and medical groups as well as the general public) in discussions surrounding the scientific and ethical issues arising from scientific developments.

The report on *Ethical, Legal and Social Issues in Human Stem Cell Research, Reproductive and Therapeutic Cloning* (Stem Cell Report)[116] issued by the BAC recommended that research on human stem cells be permitted but only in a strictly regulated manner. In addition to the observance of ethical principles (ie. informed consent), the BAC recommended prohibiting the commercialisation of donated materials (especially gametes and surplus embryos). It also recommended the establishment of a legislative and regulatory framework, with institutions empowered to license and closely monitor all human embryo research.

Furthermore, in this same report, the BAC adopted an intermediate position supporting the special status of the human embryo. The BAC specified that the human embryo is entitled to a certain degree of respect due to its potential to be a human being. However, it notes that the embryo's right is not absolute (ie. the human embryo does not have the same status as a living child or an adult) and its right may be weighed against the benefits that arise from research. It is interesting to note that the BAC's position is consistent with that of its Canadian counterpart. Indeed, the Canadian TCPS has stated that their "position recognizes ... that the present status of the law, ethics and health care in Canada regarding research ... is broadly consistent with a graduated approach that correlates permitted interventions with the developmental stages of the human embryo or foetus".[117]

---

[115]The Bioethics Advisory Committee's Consultation Paper on *Human Stem Cell Research*, 2001, was followed by two consultation processes (and a subsequent report) addressing the donation of human eggs for research and the use of human-animal combinations in biomedical research. These two consultation processes and ensuing recommendations are aimed at revising the Stem Cell Report of 2002. See the Consultation Paper on *Donation of Human Eggs for Research*, 7 November 2007, and the Report of the same title issued on 3 November 2008. See also the Consultation Paper on *Human-Animal Combinations for Biomedical Research*, 8 January 2008.

[116]Bioethics Advisory Committee, Singapore, *Ethical, Legal and Social Issues in Human Stem Cell Research, Reproductive and Therapeutic Cloning*, June 2002.

[117]Canadian Institutes of Health Research, Natural Sciences and Engineering Research Council of Canada, Social Sciences and Humanities Research Council of Canada, *Tri-Council Policy Statement: Ethical Conduct for Research Involving Humans*, 1998 (with 2000, 2002, 2005 amendments), Section 9, p. 9.1.

Additionally, the BAC recommended banning human reproductive cloning while supporting therapeutic cloning, provided that there is strong scientific merit in, and potential medical benefit from, such research. Where the derivation and use of human embryonic stem cells are permitted, these cells should first be taken from existing cell lines, originating from embryos less than 14 days old. If existing cell lines are not available, surplus human embryos that have been created for fertility treatment less than 14 days old can then be used. Finally, the BAC acknowledged the apprehension that the creation of embryos by cloning techniques may lead to the cloning of a whole human being and states that the only way to prevent this is by banning the implantation of any cloned embryo into a womb.

In conformity with the BAC recommendations, the Government of Singapore took two important actions. First, it enacted the *Human Cloning and Other Prohibited Practices Act* of 2004 (Human Cloning Act).[118] The *Human Cloning Act* prohibits placing a human embryo clone in the body of a human or the body of an animal. It also prohibits the import and export of any human embryo clone into and out of Singapore, as well as the commercial trading of human eggs, sperm, and embryos. Moreover, the Human Cloning Act also bans certain practices associated with reproductive cloning activities such as developing human embryos created other than by fertilisation of human egg by human sperm for a period of more than 14 days, prohibiting the development of human embryo outside the body of a woman for more than 14 days. In addition, the removal of human embryos from the body of the woman for the purpose of collecting a viable human embryo is forbidden.

Second, the Singapore Government endorsed the BAC recommendations pertaining to stem cell research and in consequence researchers in Singapore now have to strictly adhere to the recommendations set out in the Stem Cell Report of 2002. The absence of "unduly onerous political or legislative restrictions in hESC [human embryonic stem cell] research."[119] has created a prosperous environment allowing stem cell research to flourish in Singapore and consequently positioning the country at the forefront of the field.

The policy framework established by the BAC is reinforced by legislation regulating the activities of fertility centres in the country. The *Directives*

[118] Singapore Statutes: *Human Cloning and Other Prohibited Practices Act* (Cap. 131B), Revised 2005.

[119] Colman A. Stem Cell Research in Singapore, *Cell* **132** (2008): 519–521, p. 520.

for Private Healthcare Institutions providing Assisted Reproduction Services (Assisted Reproduction Directives) issued by the Ministry of Health[120] established the regulatory framework for all research involving human embryos and oocytes. The *Assisted Reproduction Directives* implicitly allow the creation of human embryos for research purposes, including the creation by SCNT and *in vitro* fertilisation of human oocytes with human sperm. It further allows for the donation of oocytes from healthy volunteers (women who are not undergoing fertility treatment). In addition, the guidelines set up licensing, ethics review and other governance and ethical requirements (informed consent, prohibition of commercialisation, measures to avoid conflict of interest, etc.).

Singapore's prospective approach to biomedical regulation is evident in the BAC's recent endeavours directed towards reviewing its *Stem Cell Report* of 2002. These endeavours seek to address the socio-ethical and legal implications arising from scientific and technical advances in the field by focusing on two specific issues:

1. Donation of human eggs for research; and
2. Use of human-animal combinations for biomedical research.

First, in late 2007, the BAC conducted a public consultation addressing the ethical implications of donating human eggs (oocytes) for research. Today, Singapore's regulatory framework allows healthy volunteers to donate eggs for research, provided that the governance and ethical safeguards established in the *Assisted Reproduction Directives* are complied with.[121] However, there is no legislation addressing the possibility of compensating donors for time, risk, and inconvenience while preventing undue inducement. To be sure, the *Human Cloning Act* explicitly prohibits the commercial trading of human eggs and embryos, but does not deal with the issue of providing financial incentives or monetary payments for donation of human gametes. The resulting report adopted by the BAC provided seven recommendations dealing

[120] Ministry of Health, Singapore, *Directives for Private Healthcare Institutions Providing Assisted Reproduction Services: Regulation 4 of the Private Hospitals and Medical Clinics Regulations* (Cap. 248, Reg. 1), September 2001; Revised March 2006.

[121] Ministry of Health, Singapore, *Directives for Private Healthcare Institutions Providing Assisted Reproduction Services: Regulation 4 of the Private Hospitals and Medical Clinics Regulations* (Cap. 248, Reg. 1). (September 2001, Revised March 2006).

with informed consent, compensation and care of donors, the import and use of eggs for research as well as the need for regulatory control. The recommendations are consistent with international and national norms on the matter.[122]

Second, the public consultation process launched by the BAC on *Human-Animal Combinations for Biomedical Research*[123] sought public and professional opinion on the permissibility of creating and using human-animal combinations (eg. human-animal chimeras and hybrids) in research, and the necessary regulatory mechanisms that should apply to such practice. Aiming at revising Singapore's policy framework for biomedical research (specifically for stem cell research), the results of the public consultation and the final recommendations to be adopted by BAC will provide a wide-ranging framework for this type of research. Currently, there are no specific legislative provisions or regulations addressing research involving chimeras and hybrids. However, there is provision for a ban on creating true hybrids (combinations of eggs and sperm from human and animals) for reproductive purposes under the *Assisted Reproduction Directives*.

As we can elucidate from the preceding policy review, Singapore endorses a liberal approach to genetics and stem cell research. Its approach to policy-making — a combination of public and private ordering — reflects the country's clear commitment to scientific and medical progress on the basis of their potential benefits for humanity. However, this commitment is not unfettered since, through legislative action and self-regulation, Singapore has created a solid research governance framework which aims to protect the rights, interests, and welfare of patients, research participants, and the public. The challenge ahead is to continue to provide a coherent, transparent, flexible yet enforceable system, given their mixed approach to policy making.[124]

---

[122] Isasi RM and Knoppers BM. Monetary Payments for the Procurement of Oocytes for Stem Cell Research: In Search of Ethical and Political Consistency, *Stem Cell Research* **1** (2007): 37–44.
[123] Bioethics Advisory Committee, Singapore, *Human-Animal Combinations for Biomedical Research: A Consultation Paper,* 8 January 2008.
[124] Isasi RM and Knoppers BM. Beyond the Permissibility of Embryonic and Stem Cell Research: Substantive Requirements and Procedural Safeguards, *Human Reproduction* **21**, 10 (2006): 2474–2481.

CONCLUSION: TRENDS

Since 1990, international normative frameworks have been providing guidance for emerging research in genetics. While the early 1990s understandably treated the socio-ethical and legal issues with some caution, the completion of the human genome mapping efforts in 2003 led to an interest in understanding both the role of genetic factors in common diseases and the role of gene-environment interactions via population studies. Certain research areas, such as stem cell research moved from a 'reactionary' approach (the result of an over-emphasis on embryonic stem cell research and human reproductive cloning) to addressing the potential of stem cells coming from other sources (ie. umbilical cords, adult stem cells, etc.) and concerns over their translation into therapeutic or clinical applications. Moreover, the building of international research consortia in both genetics and stem cell research is forcing the further elaboration of international self-regulatory policy-making.

In the area of large population studies, the Public Population Project in Genomics (P^3G) founded in 2007 has developed not only the scientific tools for international interoperability between 30 large population banks but also adopted a *Charter of Principles* and provided practical advice on governance, consent, and access mechanisms.[125] Building on this approach, the International Cancer Genome Consortium has built a virtual federated database with policy guidance for participating countries.[126]

Likewise, in the field of stem cell research, the emergence of national stem cell banks is accompanied by the establishment of international initiatives addressing harmonisation, cooperation, and standardisation processes for such research and banking.[127] These initiatives share the vision of stem cell research as a global enterprise and aim for the timely realisation of the scientific promise offered via the International Stem Cell Banking Initiative of the ISCF, the European Commission's Human Embryonic Stem Cell Registry (hESCReg), and the ISSCR.[128]

---

[125] Information on Public Population Project in Genomics (P^3G) is available at: http://www.p3g.org.

[126] Information on International Cancer Genome Consortium is available at: http://www.icgc.org.

[127] Knoppers BM. Reflections: The Challenges of Biotechnology and Public Policy, *McGill Law Journal* **45** (2000): 559–566; and Knoppers BM and Le Bris S. Genetic Choices: A Paradigm for Prospective International Ethics? *Politics and Life Sciences* **13**, 2 (1994): 228.

[128] Isasi RM and Knoppers BM. Governing Stem Cell Banks and Registries: Emerging Issues, *Stem Cell Research* **3** (2009): 96–105.

Building on the international principles established by international governmental (UN, UNESCO, WHO, CIOMS) and non-governmental (HUGO) bodies, these two initiatives (ie. professional self-regulation together with normative guidance from international bodies) are translating and expanding the principles into interoperable policies for international research endeavours. As such, they typify the post-genomic tendency to move across disciplines and jurisdictional borders to a more complex systems approach to policy-making.

8

Public Engagement and Bioethics Commissions

Thomas H Murray and Ross S White

A HISTORY OF BIOETHICS COMMISSIONS IN THE US

With the ever increasing importance of medicine and science in the daily lives of individuals across the globe, it is no surprise that the field of bioethics has grown over the last few decades. With this growth has come a proliferation of national and transnational commissions composed of experts in science, medicine, religion, law, and other disciplines who are called upon to weigh in on complex issues of moral significance to citizens. Although many early commissions arose in response to a specific event, case, or problem that needed attention, bioethics commissions are now routinely established in many countries, including much of Europe, the US, and other nations with a significant commitment to biomedical science such as Singapore. In order to better understand the function and origins of bioethics commissions, a brief examination of the history of such commissions in the US may be helpful.

Although the US had what were called 'blue ribbon commissions' as far back as the 18th century, the first national bioethics body was established in 1974 by the US Congress.[1] The National Commission for the Protection of Human Subjects of Biomedical and Behavioral Research (National Commission) was created "to identify the ethical principles of human subjects research and to recommend federal policy for research".[2] The work of this first

[1] Meslin EM and Johnson S. National Bioethics Commissions and Research Ethics, in *The Oxford Textbook of Clinical Research Ethics*, eds. EJ Emanuel *et al.* New York: Oxford University Press, 2008.

[2] US Congress, *National Research Act*, 12 July 1974, Public Law 93-348, Title H, Part A.

bioethics commission profoundly influenced both political and public opinion on medical practice and research. In 1979, the Commission published its influential work, the *Belmont Report,* which proposed a set of ethical principles meant to inform research involving humans. That report is regularly referenced to this day. The Commission issued a series of reports on research ethics in a range of situations and populations. It recommended the creation of committees to review proposed research and protect the human subjects of such research. The current system of Institutional Review Boards, or IRBs, was given great impetus by the Commission's analyses and recommendations. Since then, six additional time-limited national bioethics commissions have been established in the United States.

The National Commission was dissolved in 1979 and overlapped with the creation of the Ethics Advisory Board (EAB) in 1978. The EAB was charged with exploring issues surrounding research and the use of human embryos. Although the EAB recommended the use of federal funding for *in vitro* fertilisation (IVF) research and concluded that "... the human embryo is entitled to profound respect, but this respect does not necessarily encompass the full legal and moral rights attributed to persons",[3] the ethical acceptability of IVF research to the broader American public was largely unresolved and a 15-year moratorium on federal funding for IVF research ensued.[4]

The third commission, the President's Commission for the Study of Ethical Problems in Medicine and Biomedical and Behavioral Research, was created by Congress in 1978 at the request of President Carter. The President's Commission completed its mandate by the end of 1982, issuing final reports on ten pressing issues: the definition of death; informed consent; genetic screening and counselling; differences in the availability of health care; life-sustaining treatment; privacy and confidentiality; genetic engineering; compensation for injured subjects; whistleblowing in research; and an IRB guidebook. Although the Commission, led by Alex Capron, was remarkably productive, sceptics

[3] Ethics Advisory Board, US Department of Health Education and Welfare, *Report and Conclusions: HEW Support of Research Involving Human In Vitro Fertilization and Embryo Transfer,* 4 May 1979, Conclusions 1 and 2, pp. 101, 104–107.

[4] Irving DN. What is 'Bioethics'? in *Life and Learning X: Proceedings of the Tenth University Faculty for Life Conference,* ed. JW Koterski. Washington, DC: University Faculty for Life, 2002, pp. 1–84.

raised concerns about its makeup and the chance that "speculations of the ethicists become the law of the land".[5]

Though born under a Democratic administration, by the time it took on the politically controversial issue of justice and access to health care almost all of the original commissioners had been replaced by members appointed by the Republican administration under President Ronald Reagan. For many observers, this political transformation made all the more remarkable the conclusion in its report, *Securing Access to Health Care: The Ethical Implications of Differences in the Availability of Health Services*, that there is an "ethical obligation on the part of society to ensure that all Americans have access to an adequate level of care without the imposition of excessive burdens".[6] It has taken nearly 30 years for this ethical obligation to be translated into law.

The President's Commission was followed by the Biomedical Ethics Advisory Committee, which had a short and unproductive life from late 1988 to early 1990. During that time it held only two meetings. The Committee was unable to make any progress or generate substantive reports as its parent group (the Biomedical Ethics Board composed of six senators and six members of Congress, equally divided between Republicans and Democrats) became politically deadlocked over abortion politics.[7]

In 1995, President Clinton announced the creation of a National Bioethics Advisory Commission (NBAC). The Commission was called upon to explore issues beyond IVF, foetal, and clinical research, which had been a central focus of many previous bioethics commissions. The Commission's primary responsibility was to make recommendations to the National Science and Technology Council and other appropriate government entities. Through its five-year tenure, the Commission published reports on cloning human beings, research involving individuals with mental disorders that affect decision-making capacity, research involving human biological materials, human stem cell research, ethical issues related to international research, and issues related to research on human participants. These reports evoked a variety

---

[5]Jonsen AR. *The Birth of Bioethics*. New York: Oxford University Press, 1998, pp. 109–110.
[6]The President's Commission for the Study of Ethical Problems in Medicine and Biomedical and Behavioral Research, *Securing Access to Health Care: The Ethical Implications of Differences in the Availability of Health Services*, March 1983.
[7]Irving DN. What is 'Bioethics'? in *Life and Learning X: Proceedings of the Tenth University Faculty for Life Conference*, ed. JW Koterski. Washington, DC: University Faculty for Life, 2002, pp. 1–84.

of responses from the President, Congress, federal and state governments, professional societies, advocacy groups, other countries, and international organisations.[8] Soon after President George W Bush was inaugurated in January 2001, this Commission's charter was allowed to expire.

President Bush then created the President's Council on Bioethics. While this Commission was able to publish several reports before being dissolved in 2009, it was met by harsh criticism from its very beginning. Critics have often accused the Council of being too ideologically conservative and publishing reports that largely support the viewpoints of the administration that created it.[9] In fairness, it must be noted that the Council's leadership claims that its membership was more politically diverse than its predecessors, a claim that remains in dispute. During its tenure, the Council published reports on a range of issues: human cloning, considerations of dignity in emerging biotechnologies, stem cell research, ethical concerns with new reproductive technologies, newborn screening, issues faced in caring for an aging population, and the determination of death.[10] The long-term impact of these reports is difficult to assess at this early date, but there is no doubt that the President's Council thrust itself into the centre of national political and moral debates with far greater savvy and determination than previous bodies.

In November 2009, President Barack Obama announced the creation of the Presidential Commission for the Study of Bioethical Issues. The recently appointed 13-member commission has been called upon to "advise the President on bioethical issues that may emerge as a consequence of advances in biomedicine and related areas of science and technology".[11] The exact issues that the Commission will explore are still not clear, but its membership encompassing a wide range of expertise and viewpoints, including six scientists and two lawyers, suggests it could explore a variety of emerging issues. The new Commission's Chair, Amy Gutmann, is a highly regarded university president (University of Pennsylvania) and political philosopher. Her scholarly work on democracy and deliberation gives some hope that this new

[8] Eiseman E. *The National Bioethics Advisory Commission: Contributing to Public Policy*. Arlington, VA: RAND, 2003.
[9] Green RM. For Richer or Poorer? Evaluating the President's Council on Bioethics, *HealthCare Ethics Committee Forum* **18**, 2 (2006): 108–124.
[10] For information on former bioethics commissions in the US, see: http://www.bioethics.gov.
[11] Establishing the Presidential Commission for the Study of Bioethical Issues, *Federal Register* **74**, 228 (30 November 2009).

body may explore novel and perhaps more effective methods for engaging the American public in its work.

With a better understanding of the history of bioethics commissions in the United States, we now turn to an examination of the role and influence that these commissions have played in helping the American people and policy makers understand and respond to complex bioethical issues. Of particular interest is the extent to which these commissions have been able to fulfil their role to both advance meaningful understanding of these issues, as well as ensure that the public has been able to take part in the discourse.

THE ROLE OF BIOETHICS COMMISSIONS IN PUBLIC DISCOURSE

Bioethics commissions serve the important role of representing the interests of all members of the public, not merely a specific subset of the population. Their effectiveness as a body is thus largely contingent on their ability to show public and institutional legitimacy. For this reason, bioethicists on commissions become "house intellectuals",[12] who must jointly seek a common morality and take a special deliberative role in public discourse on issues of medicine and science. John Evans contends that this elevation of certain individuals to the status of bioethicists on a national commission lends itself to a technocratic mindset that presupposes specialised knowledge, and therefore expertise on difficult moral questions.[13]

For these reasons, the role of ethicists on bioethics commissions has at times been subject to three recurring lines of attack. The first of these is the 'ivory tower' critique, which asserts that trained philosophers use standards and methods "better suited to the halls of academe than to the halls of Congress".[14] Some ethicists are driven by norms of modern philosophy that are designed more for criticising a proposed solution than for generating a constructive proposal of their own. The second 'sceptical' critique calls into question the very possibility of an ethics 'expert'. The idea that any individual

[12] Evans JH. Between Technocracy and Democratic Legitimation: A Proposed Compromise Position for Common Morality Public Bioethics, *Journal of Medicine and Philosophy* **31** (2006): 213–234, p. 219.

[13] *Ibid.* p. 221.

[14] Weisbard AJ. The Role of Philosophers in the Public Policy Process: A View from the President's Commission, *Ethics* **97** (1987). 776–785.

can have a specialised grasp of morals presupposes that a given understanding of morality is somehow more 'right' than another. The third, ethnographic critique, set forth mainly by sociologists and anthropologists, calls for more empirical input into decision-making that acknowledges that moral decision-making is highly context-dependent.[15] Creating generalised prescriptions for the entire public can fail to appreciate the intricacies of individual beliefs and may be prone to shortsighted and vague assumptions about how best to address concerns of particular segments of the population.

Commissions must at the very least be cognisant of these three criticisms — the ivory tower critique; the scepticism about moral 'expertise'; and the 'ethnographic' claim about the preeminence of facts and context. At best, bioethics commissions can seek to ensure that they do not fall victim to these concerns. One way to gain the necessary legitimacy to make recommendations would be by better connecting the public and philosophical authority of bioethics. This would involve better dialogue between the academic and public sphere, and seeking to prevent 'ivory tower' moral prescriptions from trumping public concerns. Commissions can unknowingly forge and reinforce divisions by failing to weigh public discourse into their moral calculus and not meeting their obligation to serve wider public interests. To what extent can the public truly be a part of the decision-making process if these meetings are advertised as 'open to the public', but held at a place and time that is largely inaccessible to most?[16] Such systematically exclusionary structures and practices support the presumption that philosophical underpinnings have taken precedent over, and at times trumped, the views of the public.

In order to best meet the expectations of those who create them, as well as the public, Dzur and Levin propose that bioethics commissions be "assessed primarily as agenda-setting rather than expert bodies and be judged successful according to their capacity to facilitate a wider public dialogue over ethical issues . . . rather than . . . to find the best possible answers".[17] Bioethics commissions, they argue, should not be authoritative prescriptive bodies charged with choosing one stock answer for the most difficult questions. Effectiveness derives from commissions' ability to cultivate a two-way model of communication as both receivers and transmitters of knowledge about public concerns

[15]Dzur AW and Levin D. The 'Nation's Conscience': Assessing Bioethics Commissions as Public Forums, *Kennedy Institute of Ethics Journal* **14**, 4 (2004): 345–348.
[16]*Ibid.* p. 338.
[17]Dzur AW and Levin D. The Primacy of the Public: In Support of Bioethics Commissions as Deliberative Forums, *Kennedy Institute of Ethics Journal* **17**, 2 (2007): 133–142, p. 134.

and experiences. This is in contrast to a one-way technocratic commission that transmits knowledge to a "malleable public" or a no-way model that stymies transmission of knowledge either to or from the public that is "purely political".[18] Although commissions may be given the task of drawing up specific recommendations on particular subjects, Dzur and Levin claim that they must not be driven purely to seek that end.

Commissions can become agenda-setters that help to mold public discourse in such a way that cultivates public deliberations and allows a plurality of views to be heard. This "education in a democratic polity" may be "the primary function of the social-issue presidential commission".[19] The creation of these commissions also represents a symbolic acknowledgment that the authority establishing them is aware of and prepared to address the contentious issues at hand. While some commissions are designed to legitimise already settled policy stances, others are given the important task of wading through and recommending new policy options. Allowing public comment and debate in this process has proven to further, not impede, the authority of executives such as the President,[20] by allowing norms to be shaped in a democratic and fair way.

REPORTS OF THE US AND SINGAPORE ON HUMAN TISSUE RESEARCH

In order to get a better sense of how different countries deal with similar complex bioethical issues, we compare the reports written by the NBAC and Singapore's Bioethics Advisory Committee (BAC) on human tissue research.

Research Involving Human Biological Materials: Ethical Issues and Policy Guidance

In 1999, the NBAC published a report on research involving human biological materials.[21] In issuing its report to President Clinton, the NBAC chose to focus on the following questions: How well does the existing Federal Policy

[18] *Ibid.* p. 136.

[19] Flitner D. *The Politics of Presidential Commissions.* Dobbs Ferry, NY: Transnational Publishers, 1986, p. 5.

[20] Dzur AW and Levin D. The Primacy of the Public: In Support of Bioethics Commissions as Deliberative Forums, *Kennedy Institute of Ethics Journal* **17**, 2 (2007): 133–142, p. 140.

[21] National Bioethics Advisory Commission, USA, *Research Involving Human Biological Materials: Ethical Issues and Policy Guidance,* Vol. I, August 1999.

for the Protection of Human Subjects (the Common Rule)[22] meet the objective of protecting human subjects from harm in research involving human biological materials; and does it provide clear direction to research sponsors, investigators, IRBs, and others regarding the conduct of research using these materials in an ethical manner?[23] The NBAC concluded that while the overall structure of the federal regulations was generally adequate for addressing this area of research, the Common Rule was not entirely responsive to these questions, providing what is sometimes ambiguous guidance on issues such as how to define minimal risk for 'human subjects' that donate tissue. In order to illuminate some of the shortcomings of existing regulations, the NBAC made 23 recommendations.[24]

In one chapter, "Ethical Perspectives on the Research Use of Human Biological Materials",[25] the NBAC explores some of the foundational ethical issues that drove their recommendations. The Commission's discussion relies heavily on the three principles articulated decades ago by the National Commission in the Belmont Report which are meant to guide discussion on the ethics of research with human subjects: beneficence, respect for persons, and justice.

According to the NBAC report, "beneficence encompasses not only research efforts to produce generalizable knowledge that can benefit society, but also efforts to avoid harming persons, to minimize possible harms, and to assess possible harms in relation to possible benefits".[26] The NBAC's primary concern is in relation to non-physical harm that may arise out of the acquisition, use, or dissemination of information obtained from research samples. For this reason, they stress the importance of preserving the confidentiality of information obtained through research and ensuring that any such information cannot be used for discrimination in health insurance or employment, or lead to the harm of stigmatisation.[27]

In regards to the principle of respect for persons, the Commission states that every person has an interest in being treated as a moral agent, "capable

[22] *US Code of Federal Regulations*, Title 45, Part 46.
[23] Shapiro H, Letter to The President, 16 July 1999.
[24] *Ibid.*
[25] National Bioethics Advisory Commission, USA, *Research Involving Human Biological Materials: Ethical Issues and Policy Guidance*, August 1999, pp. 41–54 (Chapter 4).
[26] *Ibid.* p. 42.
[27] *Ibid.* pp. 42–45.

of exercising choices consistent with his or her own values, preferences, commitments, and conceptions of the good".[28] This provides the moral justification for informed consent, because it ensures that patients and research subjects are treated respectfully as agents, not merely passive objects to be used for the ends of others. Informed consent also entails treating individuals as autonomous agents worthy of respect, who have a right to know how their biological material may be used. Additionally, researchers must acknowledge limits requested by subjects or their families on the use of biological material for religious or cultural reasons, such as beliefs about the integrity of the body, whether living or dead.

Justice requires the fair and equitable distribution of benefits and burdens in research, demanding the treatment of similar cases in a similar way and being cognisant of relevant similarities and differences among individuals and groups. The Commission calls for consideration to ensure fair participation on the part of particular groups in the design and implementation of research protocols that may affect that group. Some populations may bear a greater burden in the research process, so measures must be taken to ensure that they are apprised of equitable benefits ascertained from the research. In connection with the principle of respect for persons, justice may also require considerations of economic benefits that may arise as a result of research, and whether or not participants have a right to reap any financial gains from their participation.[29]

Human Tissue Research: A Report by the Bioethics Advisory Committee, Singapore

In 2002, the Singapore BAC issued its report on human tissue research.[30] The principal objectives of the report were to review current issues affecting the conduct of human tissue banking and human tissue research in Singapore; recommend a national framework for the proper governance of research tissue banking activities in Singapore; and recommend a body of appropriate ethical principles and guidelines for the ethical conduct of research tissue banking and human tissue research in Singapore.[31] The Committee set forth

[28] *Ibid.* p. 47.
[29] *Ibid.* p. 49–50.
[30] Bioethics Advisory Committee, Singapore, *Human Tissue Research*, November 2002.
[31] *Ibid.* p. 11.

four broad recommendations: the adoption of a set of governing ethical principles in the conduct of research tissue banking; that research tissue banking be conducted only by institutions such as may be approved or licensed by the proposed statutory authority to do so, and not by private individuals or groups; that all research tissue banking should be licensed by a statutory authority with appropriate supervisory jurisdiction; and that given a rapidly shifting body of ethics, legal rules, and opinion on human tissue research and banking, a continuing professional and public dialogue should be initiated.[32] We will examine the proposed governing ethical principles here.

The report emphasises that "the health, welfare and safety of the donor shall be the paramount consideration in the taking of any tissue".[33] The Committee goes on to say that if tissue is taken primarily for therapeutic or diagnostic purposes, research considerations should not be allowed to compromise or prejudice the original purpose of the taking and a sub-sample must not be taken until the diagnostic or therapeutic purposes have been fulfilled. More broadly, tissue samples should only be taken for research if the potential benefit outweighs the potential risks to the patient. This relates directly to the subsequent principle of informed consent, requiring that tissue shall only be taken or accepted if full, free, and informed consent of the donor has been obtained. In certain cases, it may be ethically acceptable for research to proceed without consent provided that appropriate precautions, such as anonymisation, are observed. Finally, in regards to informed consent, tissue banks should develop electronic databases that enable consent status and consent conditions (if applicable) to be tracked.[34]

The Committee goes on to require that the human body and its remains be treated with respect. Researchers and tissue bankers are to be sensitive to religious and cultural perspectives and traditions that may require tissue to be handled in a particular way. To further preserve respect, researchers should be clear about the scope of the intended gift, particularly if it involves the donation of substantial parts of organs which may exceed the traditional understanding of what a tissue donation means. This is reinforced by the next guiding principle, which requires that donations of tissue samples for use in research be accepted only if they are given as outright gifts. This includes

---

[32] *Ibid.* pp. 4–6.
[33] *Ibid.* p. 33.
[34] *Ibid.* p. 34.

making it clear to the donor that they should not expect any personal or direct benefit from the donation, and are not apprised of information discovered, unless it has been agreed upon in advance of the donation.[35]

The report also recommends that all research using human tissue samples should be approved by an appropriately constituted research ethics committee or institutional review board. These committees and review boards must be formed in a transparent fashion, which considers and addresses potential conflicts of interest. Finally, researchers and all other individuals involved in the conduct of research tissue banking have an obligation to protect the confidentiality of the personal information of donors, as well as the privacy of donors. Furthermore, researchers must also protect the confidentiality of information given to them by donors about other individuals. Information, such as family history, must be afforded the same protection as that provided to donors.[36]

A COMPARISON OF THE REPORTS

By comparing the reports on human tissue written by the NBAC and Singapore's BAC, one can see a great deal of overlap in their ethical frameworks. Both reports emphasise the importance of autonomy and informed consent to the donation of human tissue, primacy of the welfare of the donor (beneficence), respect for the human body, and preservation of confidentiality. The guiding ethics and recommendations of the NBAC's report are heavily founded in a principlist approach derived from the Belmont Report. While Singapore is not as rigidly 'principlist', it certainly takes these ethical concepts into consideration in the formulation of its recommendations. One notable difference between the two reports is the way that they handle the idea of justice. While the NBAC seems to suggest that researchers should consider issues of what potential benefits donors may find from the donation of their tissue, BAC seems to suggest that by virtue of donations being 'gifts', donors are less entitled to benefit gleaned from their donation so long as it is in accordance with the consent process. While these bioethics bodies in different countries reached very similar conclusions and recommendations, the formulation of these recommendations was probably heavily influenced by the differences in the decision-making processes.

[35] *Ibid.* p. 35.
[36] *Ibid.* p. 36–37.

Another important difference concerns the entities creating and maintaining research tissue banks. The NBAC made no attempt to alter the eclectic mix of tissue-collecting organisations in the US. Tissue collections are held by hospitals, research institutes, universities, state public health departments (in the Guthrie cards collected by newborn screening programs) and, for the single largest collection, the Armed Forces Institute of Pathology, the military. Nor did the NBAC propose a uniform licensing or certification regime for all tissue banks. Singapore's BAC, in contrast, proposed the creation of a statutory authority to approve or license tissue banks, that no tissue banks be permitted to be in the hands of private individuals or groups, and that this new authority should have supervisory oversight of all tissue banks.

The difference in approaches seems to reflect at least two factors. First, although the US is a federalist system, all powers not delegated to the national government continue to reside within the 50 states. Newborn screening, for example, is conducted separately by each state, and each state makes its own determination of what conditions should be screened for, and how the Guthrie cards with their dots of dried blood will be stored, used for any other purpose, and eventually destroyed. Recent efforts spearheaded by the federal government and a well-known private charity, the March of Dimes, have resulted in more uniform screening panels, but have not had notable impact on the tissue banking policies or practices. The layered and diffused nature of political authority makes it difficult to create and enforce a single, central policy for tissue banks. Singapore, as a unified political entity, does not have to cope with jurisdictional conflicts, making a uniform national policy far easier to create.

Second, US researchers and institutions may be much more suspicious of government control. Such attitudes set the US off from many other countries, including most European nations. In this regard, Singapore may be much closer to Europe than to the US in its willingness to accept government regulation and control over tissue banks.

PUBLIC ENGAGEMENT: A FOUR-STAGE PROCESS

In order to better understand the level and role of public deliberation in bioethics commissions, we propose as an organising conceptual framework a four-stage process of engagement: membership, voices, interaction, and report. While our analysis seeks a broad-based understanding of how to fruitfully engage the public, it has a special concern for how minority views, particularly minority religious views, are solicited and incorporated into the deliberation and recommendation process. To help illuminate this interaction process we examine the public involvement in the deliberations of the NBAC and the BAC on human tissue research.

Membership

The first stage of public engagement we shall call 'Membership'. This stage involves efforts to identify and appoint appropriate members to commissions so as to represent the breadth of opinions and ideas held by the public, academics, scientists, and all other interested parties. This stage is the first avenue for public involvement and deliberation as it confers legitimacy on views expressed as a member of the commission. An individual is also ensured direct access to the decision-making process as an officially appointed member of a commission.

The composition of bioethics commissions is subject to many different restraints. As a whole, the general public is not in a position to directly decide the makeup of an ethics committee. Even then, effort should be taken to ensure that those chosen to represent the public are able to express views that the public will acknowledge as legitimate and important. Often, the best the public can hope for is representation on a micro-level — choosing members in a way that attempts to reflect the views of relevant subgroups of the population.[37] These representatives, entrusted with echoing public sentiments, are given the difficult task of attempting to articulate the views of many. Especially problematic is the struggle to ensure that minority views are indeed considered.

The NBAC aimed for this sort of micro-level representation through its charter:

> *At least one member shall be selected from each of the following categories of primary expertise: (i) philosophy/theology; (ii) social/*

[37] Friele MB. Do Committees Ru(i)n the Bio-Political Culture? On the Democratic Legitimacy of Bioethics Committees, *Bioethics* **17**, 4 (2003): 301–318, p. 310.

behavioral science; (iii) law; (iv) medicine/allied health professions; and (v) biological research. At least three members shall be selected from the general public, bringing to the Commission expertise other than that listed. The membership shall be approximately evenly balanced between scientists and non-scientists. Close attention will be given to equitable geographic distribution and to ethnic and gender representation.[38]

The BAC has a less regimented structure for membership, stating simply: "The BAC consists of thirteen members who reflect a range of legal, biomedical, academic and administrative expertise."[39] It does not have stipulated formalised provisions to ensure diversity of opinions and membership, but the actual composition of the BAC suggests that such considerations were taken into account. In fact, the 13-person body is comprised of at least one individual each from the fields of medicine, philosophy, the media, religious organisation, law, and communications. These attempts at broader representation by both the NBAC and the BAC demonstrate an interest in pursuing meaningful and representative deliberation through diversity of views.

Although striving for representative bodies is likely the easiest way to foster public engagement in the 'Membership' phase, bioethics commissions cannot rely exclusively upon this mechanism. Certain shortcomings or hasty generalisations must be avoided. For example, a feminist arguing in favour of the right to an abortion and reproductive liberty does not represent the viewpoint of all women on the issue and her justifications for her viewpoint can certainly not be generalised as *the* 'women's' standpoint. This also has wider implications because bioethics commissions cannot always claim legitimacy for their work by virtue of the fact that they have all interested parties represented on a micro-level. While the aims of a micro-level representation scheme are admirable, they can fall victim to unfair and unrepresentative generalisations of opinion. In order for these representatives to work effectively, however, they must have a clear sense of their objectives for deliberation, thus leading us to stage two.

[38]See Charter of the National Bioethics Advisory Commission: http://bioethics.georgetown.edu/nbac/about/nbaccharter.pdf.

[39]See description of the Main Committee of the Bioethics Advisory Committee: http://www.bioethics-singapore.org/.

Voices

The second stage of public engagement is 'Voices', which reflects the extent to which public voices are heard in helping to identify the important issues and to set an agenda for addressing their complexities. Whether or not the public is adequately represented on the commissions as members, they must be afforded the possibility to have their voices heard.

The NBAC was acutely aware of this need to gauge the views of the public prior to delving deep into the deliberative process. This is evidenced in the process taken in the crafting of its report on human biological materials. Early in the Commission's discussions on the report, it determined that a study of public beliefs and knowledge relating to research use of human biological materials would be useful. It states that "the opinions of members of the American public — those individuals who are neither medical researchers nor ethical experts — regarding the uses of stored human biological materials in research provide important additional information for consideration in its report and recommendations".[40] To its chagrin, the NBAC soon discovered that the rules and procedures under which it, as a national body, might initiate a formal scientific study would not permit such a study to be completed in time for its findings to be incorporated into its report. A series of public hearings, however, was another thing entirely: the NBAC had the authority to engage in a series of such hearings.

They contracted with the Center for Health Policy Studies (CHPS) to gather a selection of members of the public in order to explore their knowledge, beliefs, and feelings about a variety of issues related to human biological materials. The hearings focused on six areas of inquiry: consent and ownership; privacy and confidentiality; stigmatisation of ethnic groups; third-party concerns; sponsorship of research; and safeguards. Locations of the focus groups included Richmond; Honolulu, Hawaii; Mililani, Hawaii; San Francisco; Cleveland; Boston; and Miami. For each mini-hearing they sought a rather homogonous group of 7 to 14 participants: educated baby boomers, members of an urban neighbourhood board, members of two sub-urban neighbourhood boards, students and young adults, African-Americans,

[40]National Bioethics Advisory Commission, USA, *Research Involving Human Biological Materials: Ethical Issues and Policy Guidance*, Vol. I, August 1999, p. 77.

senior citizens, and Jewish women.[41] While these focus groups cannot be said to represent all segments of the general population, care was taken to include a diversity of age, race, affluence, education, and religious affiliation.

It is impossible to generalise the responses within the focus groups, but some conclusions can be drawn. Most people understood what constituted human tissue, but many were not aware that it could be stored and later used for research. The consent process was generally not well understood; distinguishing consent for the procedures used to obtain the tissue (such as a biopsy) from consent for its future use in research could be helpful. Participants were for the most part comfortable with the confidential use of their tissue, so long as insurance companies did not have access to findings, but there appeared to be a need for better public education on the management of sensitive medical information.

The NBAC did not find a generalised fear of stigmatisation of specific ethnic groups. If anything, sentiment leaned towards the hope that group-specific genetic research would have greater benefits for those groups than harms. Participants believed the research subjects should be able to choose whom to disclose information to, and in the absence of competence a legal guardian should be responsible for consent. Regardless of who sponsored the research, for example government or industry, participants thought that a societal benefit was likely. When asked who should be on the IRBs charged with approving research projects that utilise human biological materials, participants preferred IRB members with characteristics similar to those typically found on IRBs and thought at least one member of an IRB should be a representative from the group being researched.[42]

Public sessions such as these can play an important role in giving the public the power to help guide deliberation and decision-making. In this sense, the public becomes a part of the agenda-setting and deliberative process. An understanding of the starting point for public opinion on a complex issue such as human tissue research can help commission members better frame the necessary choices before them, leading to a debate that begins with an appreciation of public beliefs and sentiments. While acknowledging a wide range of voices is important in the beginning of the deliberative process, they

[41] National Bioethics Advisory Commission, USA, *Research Involving Human Biological Materials: Ethical Issues and Policy Guidance,* Vol. II, January 2000, p. G 3–5.
[42] *Ibid.* p. G 21–22.

are arguably more important in the actual deliberations that occur as a result of the agenda-setting.

Interaction

The third stage of public engagement is 'Interaction', where the deliberations occur, typically in the context of commission meetings. One way to increase intellectual and procedural transparency is to ensure that all meetings of the commission are open and accessible to the public. This helps to reduce the suspicion that decisions are being made behind the scenes in ways that are exclusionary and uninterested in outside views. The actual writing of reports is however exclusionary by nature; a group of individuals — Commission members, staff, or consultants draft the written document. As an antidote to this isolation, commissions often will solicit the advice and feedback of outside groups, including professional and public organisations, for review of preliminary drafts of their reports. The feedback they receive can be used to help guide further discussions and revisions until, after an extended iterative process, the final report is written and adopted.

When the BAC was developing its report on human tissue research, it first drafted a consultation paper. This paper was sent to a total of 66 religious, civic, professional, scientific, medical, and healthcare organisations on 27 February 2002. In close conjunction with this distribution, they held a press conference on 4 March 2002 inviting public participation and feedback on the process and content of the paper. A total of 37 parties responded to the requests for feedback, 33 of which had specific comments or suggestions for the Committee.[43]

The Committee made great efforts to reach out to a variety of religious groups and perspectives, ten in total, six of which gave formal responses to the consultation paper. The Graduates' Christian Fellowship supported the BAC's emphasis on the principles that the "human body and its remains are to be treated with respect"; however, this organisation emphasised that only foetuses and embryos resulting from miscarriage, not excess or aborted foetuses, should be used in research.[44] The Islamic Religious Council of Singapore stressed the need for researchers to be sensitive to religious sentiments and

[43] Bioethics Advisory Committee, Singapore, *Human Tissue Research*, November 2002, p. 7–8.
[44] *Ibid.* p. D 177.

seek the advice or consent of relevant religious authority to comply with Islamic law, such as the need to appropriately cleanse and bury human tissue in accordance with Islamic teachings.[45]

The National Council of Churches of Singapore expressed concerns about the potential of violating human dignity by transgressing against the body. They were most troubled by the inclusion of foetuses and embryos, which they perceive as full human beings, in the definition of human tissue.[46] The Catholic Medical Guild of Singapore voiced similar opposition, arguing that respect for human life must begin at the moment of conception and that it would be immoral to abort foetuses or use embryos for research.[47] The Jewish Welfare Board feared desecration of a dead body in research, which conflicts with the principle that every part of the body must be buried.[48] The Parsi Zoroastrian Association of Singapore supported the consultation paper, stating that donating parts of the body for the good of humanity is always noble and any such objections to the use of human tissue would reside at a personal, rather than religious, level.[49] There was no response from Buddhists, Taoists, Hindus, or Sikhs.

While the NBAC did not reach out specifically to religious groups to provide feedback on a preliminary draft of their report, they did seek out a variety of literature on religious perspectives and commissioned a scholarly paper on the subject. This commissioned paper provided specific information relating to the views of different religious traditions on uses of body parts within medicine. The religious traditions and perspectives represented were Jewish, Roman Catholic, Protestant, Islamic, Hindu, Buddhist, and Native American.[50] Many of the religious concerns about the use of human tissue in research were similar to those found by BAC. The author of the background paper set forth five recommendations: an emphasis on public education on human tissue banking; an informed consent process that obtains authorisation for the use of tissues that is specific, substantive, and sensitive to religious values about the body (personal and communal); patients should be viewed

[45] *Ibid.* p. D 196–8.
[46] *Ibid.* p. D 201–2.
[47] *Ibid.* p. D 237–8.
[48] *Ibid.* p. D 241.
[49] *Ibid.* p. D 251.
[50] Campbell C. Research on Human Tissue: Religious Perspectives, in *Research Involving Human Biological Materials: Ethical Issues and Policy Guidance*, Vol. II, National Bioethics Advisory Commission, USA, January 2000, pp. C 16–20.

as having dispositional authority over bodies; contributions of human tissue for the purpose of advancing scientific research and knowledge is ethically preferable to other modes of acquisition, such as sale or abandonment; and procedures for procuring and using human tissue for research should protect confidentiality and community interests. These recommendations were intended to "respect and acknowledge the sacral role of the body in religious discourse and practice".[51]

The detailed and specific concerns of all religious traditions could not be incorporated into the final report, but they provided valuable insight into the underlying religious considerations that play into decision-making and research related to human tissue. The act of soliciting advice demonstrated the Commission's seriousness about attending to a variety of different religious views, thereby providing more legitimacy to its final recommendations. A greater understanding of the disparate perspectives helps to create more sensible and culturally appropriate recommendations.

Report

The fourth phase of public engagement is the 'Report' phase, where the public is invited to engage with the final report and reflect upon recommendations set forth by the commissions. Public engagement should not stop when the commission has made its final solicitation for comments on a draft report. Even as the final report is released it must be subject to a critical evaluation from the public. This is the phase where it becomes evident what input from the public and other interested parties has been listened to and applied in a way sensitive to broadly held concerns.

The most obvious place that the public had a strong influence on the final report stage of the NBAC was in the formulation of its recommendations. This is especially evident in the sections related to obtaining informed consent, which echo the major concerns and findings of the hearings conducted prior to writing of the report. Recommendation 6 explicitly requires that informed consent to the research use of human biological materials be separate from the informed consent to clinical procedures, a direct concern of research participants. Recommendations 7 and 9 similarly address issues raised by the focus groups: making clear that refusal to consent will in no way affect

[51] *Ibid.* pp. C 12–16.

the quality of clinical care; and providing potential subjects with a "sufficient number of options to help them understand clearly the nature of the decision they are about to make".[52] Finally, Recommendation 23 addresses public concern about confidentiality and autonomy in relation to the use of medical records in tissue research.[53]

Similarly, the impact of public voices, especially those of the various religious communities, is most evident for the BAC report in Recommendation 1C, which states that the human body and its remains should be treated with respect. Its emphasis on sensitivity "to religious and cultural perspectives and traditions, especially when whole cadavers or gross organs are concerned"[54] reflects an understanding of pervasive religious concerns about the sanctity of the human body and the obligation to afford it all due respect. With this sensitivity to public concerns and perceptions, the commission has created more sensible and inclusive policy guidelines.

IMPACT OF PUBLIC DISCOURSE

There are several lessons that can be learnt from the involvement and engagement of the public in the deliberations and decision-making process on difficult bioethics issues. First, bioethics commissions can show respect for civic discourse by including community representatives in their membership. Second, commissions are more likely to be responsive to the public's concerns if they give the public a palpable voice in framing the issues to be taken up by them. Needless to say, attending to these voices also strengthens the commissions' ability to address widely held concerns, rather than narrowly conceived preoccupations of scientists, philosophers, lawyers, or similar experts. Third, national bioethics organisations derive a great deal of legitimacy from being able to engage the public in their deliberative processes. Those individuals who have a vested interest in and are affected by the policy outcomes of such deliberation lend a great deal to the process and will be more supportive of the recommendations if their insights and concerns were heard. Effectively engaging the public throughout the process

[52] National Bioethics Advisory Commission, USA, *Research Involving Human Biological Materials: Ethical Issues and Policy Guidance*, Vol. I, August 1999, pp. iv–v.
[53] *Ibid.* p. vii.
[54] Bioethics Advisory Committee, Singapore, *Human Tissue Research*, November 2002, p. 5.

will also confer on these organisations and their recommendations legitimacy. This legitimacy depends on "the ability of the policy-maker to justify those policies to any reasonable member of the society".[55] This explanation is much easier if the public was called upon to take part in the development of the policy in the first place. Finally, commissions or bioethics organisations should continuously engage with the broader public after their analysis and recommendations are published. Commission reports cannot be seen as the final, definitive word. Rather, they are documents that can help to frame issues that interweave complex scientific and ethical elements. A useful report is one that spurs public interest, informs in a constructive way the ongoing debate, and leads in the end to policies based on our best understandings of the relevant science, ethics, and law, as well as the pertinent beliefs of the broader community.

[55] Dodds S and Thomson C. Bioethics and Democracy: Competing Roles of National Bioethics Organisations, *Bioethics* **20**, 6 (2006): 326–338, p. 332.

9

Norm-making on Human-Animal Chimeras and Hybrids in Singapore, the United Kingdom and the International Domain

W Calvin Ho and Martin Bobrow

In Singapore, chimeras and hybrids became a more prominent subject matter in the deliberations of the Bioethics Advisory Committee (BAC) when researchers asked the Agency for Science, Technology and Research, Singapore's principal public funder of biomedical research, about the possibility of creating human-animal combinations, such as cytoplasmic hybrid (or cybrid) embryos for research. This occurred shortly after researchers in Britain proposed the creation of these embryonic constructs to compensate for the shortage of human eggs for research in therapeutic cloning or somatic cell nuclear transfer (SCNT). In November 2006, the UK Human Fertilisation and Embryology Authority (HFEA) received two applications to derive stem cells from embryos created through SCNT using animal eggs. Ethical concerns over the mixing of human with non-human genetic material arose, as the creation of cybrids involves the introduction of a complete copy of the human chromosome complement into an animal egg. Moreover, since the objective was eventually to derive embryonic stem cells, ethical debates relating to the moral status of an embryo were again aroused.

The HFEA only regulates human embryos, and it is questionable if these constructs should be regarded as 'human embryos'. If one is to understand a human embryo as arising only through the fertilisation of a human egg with a human sperm, then a cytoplasmic hybrid embryo would not technically be considered a human embryo. But if it is a question of its potentiality in developing into a human being, a cytoplasmic hybrid embryo could be regarded as a human embryo. It was on the latter rationale that the HFEA considered research involving the creation of cytoplasmic hybrid embryos to fall within its regulatory purview as it assessed them to be substantially human in that, apart from animal mitochondrial DNA in the egg cytoplasm, the bulk of the DNA would be of human origin.[1]

Cybrids and other types of human-animal combinations have been a subject of ethical inquiry in a number of countries, including Singapore. In 2008, the BAC carried out a public consultation with a view to provide recommendations to the government on the subject. While the circumstances, contexts, and rationales within which these national developments occurred were relatively distinct, a shared normative understanding of human-animal combinations like cybrids did seem to have emerged. Arguably, this phenomenon could be attributed to interaction between local and transnational policies, such as those of the International Society for Stem Cell Research (ISSCR). This chapter examines the construction of human-animal combinations as social metaphors which, as a form of cognitive order, draw normative content within, and thereby are the result of, particular socio-historical settings.[2] On normative content, this chapter further considers what Terence Halliday identifies as the three interacting cycles of norm-making: global norm-making, national norm- or law-making, and the interactive dynamics between the two.[3] In the production of this normative 'script' for human-animal combinations, recent developments in the UK and in Singapore, and the manner in which these developments relate to the policies of the ISSCR, are discussed. It is further considered whether the normative shaping occurred through recursivity or some form of cyclicality, as well as the extent to which this 'script' is likely to address inherent contradictions and indeterminacies. A broader agenda that is encompassed in these evaluations is a conceptualisation of the relationship between science, technology, and public policy on human-animal combinations.

[1] Human Fertilisation and Embryology Authority, UK, Press Statement, 11 January 2007.

[2] Lin M. *Certainty as Social Metaphor: The Social and Historical Production of Certainty in China and the West.* Westport, CT, and London: Greenwood Press, 2001.

[3] Halliday TC. Recursivity of Global Normmaking: A Sociolegal Agenda, *Annual Review of Law and Social Sciences* **5** (2009): 263–289, p. 263. Halliday defines 'norms' as "formalized codifications of behavioral prescriptions that are accepted by subjects [ie. disciplines] as legitimate and authoritative" (p. 268). This definition is adopted in this chapter.

We first consider the ways in which certain chimeras and hybrids are constituted as ethical and regulatory 'objects' through the intermediation of the BAC. It is argued that the 2008 consultation paper was a critical cognitive device in facilitating public discussion on the subject. Developments on the subject in the UK are then considered, essentially from the standpoint of the UK Academy of Medical Sciences (AMS). An important difference in the UK from the situation in Singapore is the existence of a legal framework from which certain types of human-animal combinations, such as cybrids, could draw reference. In spite of these differences, a general 'script' or shared norms on the use of human-animal combinations in stem cell research may be said to have emerged. These norms in turn reflect a broadening of the scientific agenda and could themselves facilitate cross-border research collaborations.

DEFINING HUMAN-ANIMAL COMBINATIONS IN SINGAPORE

The ethical framework for stem cell research in Singapore was established by the BAC in its 2002 report (Stem Cell Report).[4] Although it enabled stem cell research, it did not resolve a fundamental material constraint in SCNT — a technological key to unlocking the secrets to regenerative medicine. Scientists interested in SCNT need human eggs, but eggs are in limited supply since the retrieval process is invasive and poses some risk to the woman. Furthermore, a woman is unlikely to want to undergo this process unless motivated by the prospect of having children. Eggs retrieved in fertility treatment are normally fertilised for implantation or stored for future reproductive use, and will not be available for SCNT . With intensified international media coverage of the scandal revolving around Professor Hwang Woo-Suk from 2004,[5] egg donation became an issue that required policy attention. For countries supportive of SCNT , it provided a basis for further consideration as to whether there are any ethically appropriate means to increase the supply of human eggs for research or if there are any alternatives. The ethical, legal, and social issues relating to egg donation was considered by the BAC, whose recommendations on the subject were published in a report after a two-month

[4]Bioethics Advisory Committee, Singapore, *Ethical, Legal and Social Issues in Human Stem Cell Research, Reproductive and Therapeutic Cloning,* 21 June 2002.
[5]Cyranoski D. Korea's Stem-Cell Stars Dogged by Suspicion of Ethical Breach, *Nature* **429**, 6987 (2004): 3.

public consultation from 7 November 2007.[6] The creation of cytoplasmic hybrid embryos was subsequently considered under the broader rubric of human-animal combinations in a consultation paper (HA Consultation Paper) that was released to the public on 8 January 2008.[7] These deliberations and consultations on egg donation and human-animal combinations have been carried out with the intent of updating the BAC's 2002 report on stem cell research and cloning in the light of recent developments.

The overall format of the HA Consultation Paper did not differ substantially from earlier consultation papers prepared by the BAC in that there was relatively clear segmentation of discussion relating to the scientific, ethical, and legal implications of the subjects. But unlike its predecessors, the HA Consultation Paper did not propose any recommendation for consideration. The overall tone of the Consultation Paper was task-oriented, and hence pragmatic. In devising a definition for chimeras and hybrids, it did not attempt to explain the essence of 'humanity'. Instead, the focus fell on the types of human-animal combinations that were already used in research or in medical therapy and those that researchers have an interest in developing. The action orientation of policy documents required the outlay of information to be made in a way that can direct meaningful responses to policy challenges.

In fact, prior to the release of the HA Consultation Paper, human-animal combination was already a moot topic in the public domain. For instance, a commentary specifically on the subject was published in the mainstream local Chinese newspaper on 1 July 2007.[8] The commentary was entitled '人面兽身', which suggests a tendency for a lay person to associate research involving human-animal combinations with the creation of monsters.[9] In Chinese culture, unions between man and beast tend to have very bad connotations.[10] However, the newspaper commentator — a biology graduate student — indicated that most people he knew did not object to the research after the objectives and nature of these scientific constructs were clearly explained to them.

[6] Bioethics Advisory Committee, Singapore, *Donation of Human Eggs for Research*, 3 November 2008.
[7] Bioethics Advisory Committee, Singapore, *Human-Animal Combinations for Biomedical Research: A Consultation Paper*, 8 January 2008.
[8] Fieldnotes, 21 August 2008 (Ho WC).
[9] 陈华彪. 人面兽身, 联合早报 [Chen HB. Human Face, Beast Body, *Lianhe Zaobao*], 1 July 2007.
[10] Interview with sociologist and BAC member, Professor Eddie Kuo, 28 April 2009.

Clarity in defining 'human-animal combinations' was necessary in drafting the HA Consultation Paper. Purely from the standpoint of scientific capability, a broad spectrum of human-animal combinations can be produced through some combination of cells and/or genetic material. However, the consequences of the mixing are less clear. For instance, it is not known if a mouse that has a brain composed entirely of human neural cells would exhibit characteristics that one could recognise as 'human'. Even then, there are good reasons to expect that a primate (being 'closer' to human beings in evolutionary terms) is more likely to exhibit 'human' characteristics if its brain (being structurally similar to that of a human) was composed entirely of human neurons than would a mouse. In addition, the developmental stage of the research entity (be it an embryo or a fully developed organism) is relevant as this may have an effect on the extent of integration between host tissues and the introduced cells or genetic material. Hence, despite the uncertainties, useful analytical measures have been identified as the extensiveness of the human-animal mix, the type of organism concerned, and the developmental stage of this organism. A general taxonomy comprising three main types of human-animal combination, namely chimeras, hybrids, and transgenic organisms,[11] could be devised from a review of scientific, ethical, legal, and policy literatures on the subject. Table 1 sets out the possibilities categorically.[12]

However, the scope of the HA Consultation Paper is much narrower in order to aid comprehension and facilitate discussion. In narrowing down the possibilities, all categories of human-animal combinations that either did not draw any scientific interest or were ethically less controversial were excluded. Based on the latter rationale, transgenic animals are not included in the HA Consultation Paper. As for transgenic humans, there is no known scientific interest in such experimentation. Transgenic animals that are routinely used

[11] In the HA Consultation Paper, a chimera is defined as an "organism whose body contains cells from another organism of the same or a different species", whereas a hybrid is an "organism whose cells contain genetic material from organisms of different species". A transgenic animal is an "animal that has a genome containing genes from another species". See Bioethics Advisory Committee, Singapore, *Human-Animal Combinations for Biomedical Research: A Consultation Paper*, 8 January 2008, pp. 40–42.

[12] Important sources of information for this table include documents of the US National Academy of Sciences and the UK Academy of Medical Sciences. See National Academy of Sciences, USA, *Guidelines for Human Embryonic Stem Cell Research*, 2005 (amended 2007, 2008, and 2010); and Academy of Medical Sciences, UK, *Inter-specieInter-species Embryos: A Report by the Academy of Medical Sciences*, June 2007.

Table 1: Types of Human-Animal Combination

Type of human-animal combination	Definitions		Examples of uses in research
	Embryo	Developed entity	
Human chimera (with some animal cellular material)	An embryo created by introducing one or more animal cells, usually stem cells, into a human embryo at an early stage of development. Any particular cell from the resulting embryo can be traced back either to the human or animal source.	An entity brought to term from a preponderantly human chimeric embryo. Animal cells would be present in many or all of its tissues. This entity may also be created by introducing animal cells into a person or into a human embryo at a later stage of development. Such an entity would have animal cells present only in a few tissues.	No known proposals to create such chimeric embryos or entities for research purposes (other than for clinical transplant research).

(*Continued*)

Type of human-animal combination	Definitions		Examples of uses in research
	Embryo	Developed entity	
Animal chimera (with some human cellular material)	An embryo created by introducing one or more human cells, usually stem cells, into an animal embryo at an early stage of development. Any particular cell from the resulting embryo can be traced back either to the human or animal source.	An entity brought to term from a preponderantly animal chimeric embryo. Human cells would be present in many or all of its tissues. This entity may also be created by introducing human cells into an animal or into an animal embryo at a later stage of development. Such an entity would have human cells present only in a few tissues.	Growing human organs in animals for the purpose of transplantation into humans. For testing the pluripotence of stem cells, for example, via the transplantation of such cells into immuno-deficient mice. To evaluate the potential usefulness and safety of transplanting human stem cells for clinical treatment, by testing such stem cells in animals. For creating disease-specific research models such as the HIV-infected SCID-Hu mouse.

(*Continued*)

Type of human-animal combination	Definitions		Examples of uses in research
	Embryo	Developed entity	
Human-animal cytoplasmic hybrid or cybrid (with a human nuclear genome)	An embryo created by replacing the nucleus of an animal egg with the nucleus of an adult human somatic cell.	An entity brought to term from a human-animal cytoplasmic hybrid embryo with a human nuclear genome.	The embryo may be used for studying the processes involved in nuclear reprogramming, which may lead to deriving patient-specific stem cell lines. May also potentially be used to derive disease-specific stem cell lines for the purpose of research on specific diseases. No known proposals to bring a human-animal cytoplasmic hybrid embryo to term for research purposes.
Animal-human cytoplasmic hybrid (with an animal nuclear genome)	An embryo created by replacing the nucleus of a human egg with the nucleus of an adult animal somatic cell.	An entity brought to term from a human-animal cytoplasmic hybrid embryo with an animal nuclear genome.	No known proposals to create animal-human cytoplasmic hybrid embryos or entities for research.

(*Continued*)

Type of human-animal combination	Definitions		Examples of uses in research
	Embryo	Developed entity	
True human-animal hybrid	An embryo created by fertilising an animal egg with a human sperm or *vice versa*. Any particular cell from the resulting embryo contains almost equal genetic contributions from both human and animal sources.	An entity developed from a human-animal true hybrid embryo.	Not used in research, although the human-animal true hybrid embryo has been used to assess or diagnosis sub-fertility.
Transgenic human (with some animal genetic material)	A human embryo with animal genes inserted.	A human being developed from a human embryo that has integrated animal genes.	No known proposals to create such transgenic human embryos or human beings.
Transgenic animal (with some human genetic material)	An animal embryo with human genes inserted.	An animal developed from an animal embryo that has integrated human genes.	Transgenic animals with human genetic material are widely used as disease-specific research models and many examples exist. One such example is the OncoMouse.

in research tend to carry a very small number of human genes and are hence not considered to be ethically controversial. However, transgenic animals could be a matter for future deliberation if whole human chromosomes are incorporated into non-human animals. As a matter of public policy, it would not be necessary to consider all the types of human-animal combination that can be created, even if there may be academic reasons to do so. Even in the broader deliberation of the AMS, the contingency of free-roaming human-animal creatures was precluded as the subject matter was mainly confined to embryos.

Apart from narrowing the scope through the categorical exclusion of certain human-animal combinations, a processual limitation has also been adopted in that only 'human-to-animal' combinations fall within its purview. A chimera or hybrid could arise through the incorporation of animal material into humans ('animal-to-human' or 'human' chimera), or through the incorporation of human material into animals ('human-to-animal', or 'animal' chimera or hybrid). The focus would only be on the latter process as the former was either ethically unambiguous at that time or already captured within an existing regulatory framework. For instance, the creation of an animal cytoplasmic hybrid (by introducing an animal nucleus into an enucleated human egg) would be unethical as the scarcity of human eggs implies that eggs should not be 'wasted' unless there is an overwhelming scientific imperative. As for any research on human embryos, specific approval from the Ministry of Health (MOH) is required. Hence, any attempt to create a human embryo with non-human material could only be done with the approval of the Ministry. Similarly, the incorporation of animal material into humans at any point from the foetal stage of development would be regulated as research involving human subjects. With the successive narrowing of focus and the exclusion of theoretical possibilities from current consideration, the broad scope of 'human-animal combinations' was cropped down to 'human-to-animal' chimeras (or animal chimera) and cytoplasmic hybrids.

The much narrower scope is apparent in the table on the types of human-animal combination set out in the HA Consultation Paper (see Table 2).[13] Although 'transgenic animals' are included in the table, they have not been

[13] Bioethics Advisory Committee, Singapore, *Human-Animal Combinations for Biomedical Research: A Consultation Paper*, 8 January 2008, p. 15.

Table 2: Creation and Use of Chimeras and Cytoplasmic Hybrids

Type of human-animal combination	How it is created	Examples of use in research
Animal chimeras	By introducing human cells, usually stem cells, into an animal or an early animal embryo or an animal foetus.	Testing the developmental potential of human stem cells or their derivatives. Evaluating the potential usefulness and safety of transplanting human stem cells for clinical treatment. *In vivo* drug testing giving an approximation to human responses. Studying the possibility of growing human tissues and organs in animals for the purpose of transplantation into humans.
Cytoplasmic hybrid embryos	By the transfer of the nucleus of a human somatic cell into an animal egg from which the nucleus has been removed (see Figure 2 [of HA Consultation Paper]).	A source of pluripotent stem cells for research. Studying the processes involved in nuclear reprogramming. A source of disease-specific stem cells for the study of specific disease processes and methods of treatment.
Transgenic animals	By introducing human genes into an animal embryo.	Routinely used in research to understand the cause of diseases, to develop more effective treatment for these diseases, to test the safety of new products and vaccines, and to study the possibility of producing organs for transplantation that will not be rejected.

considered in the HA Consultation Paper, and serve only to emphasise that they are not considered to "raise any new ethical difficulties".[14]

Given the narrower focus, the HA Consultation Paper could have been re-titled 'chimeras and hybrids'. However, a generic expression like 'human-animal combinations' is arguably more neutral than an explicit reference to 'chimeras' and 'hybrids'.[15] In addition, if the draft consultation paper was to be renamed 'chimeras and hybrids', it could be confused with the recently concluded public consultation of the HFEA. Hence, in spite of the narrower focus of the draft consultation paper, the title of 'human-animal combinations' was used. A further possibility was for the term 'inter-species cell transplantation' to be used in place of 'chimera', since the latter was regarded as emotionally charged and carried negative connotations.[16] However, such a terminological substitution might be perceived as an attempt to sidestep ethical controversy by using a different label for something generally understood as referring to chimeras. Given that the term 'chimera' has already been used in a variety of literature to refer to the mixing of human and animal biological material at a cellular level, the terminology was used.

ETHICAL EVALUATION IN THE HA CONSULTATION PAPER

The ethical discussion in the draft consultation paper was directed at refuting what Leon Kass — Chairman of President George W Bush's Council on Bioethics from 2002 to 2005 — regarded as the 'wisdom of repugnance'. Kass argues that we shudder at the prospect of human cloning not because of the novelty of the technology, but because "we intuit and feel, immediately and without argument, the violation of things that we rightfully hold dear".[17] If cloning causes one to shudder, the prospect of human-animal chimeras and hybrids will quite possibly create convulsions. Some scholars provide convincing reasons not to dismiss feelings lightly, even if they are merely initial reactions. For example, following Martin Heidegger and Maurice Merleau-Ponty, Kym Maclaren argues that emotions are not located in some "solipsistic consciousness" but in our embodied engagement with the world and with

[14] *Ibid.* p. 13, para 14.
[15] Fieldnotes, 21 August 2007 (Ho WC).
[16] Fieldnotes, 14 November 2007 (Ho WC).
[17] Kass LR. The Wisdom of Repugnance, *New Republic* **216**, 22 (2 June 1997): 17–26.

others.[18] In addressing this reaction, a moralistic attitude was not adopted in the HA Consultation Paper as this would come across as fundamentally dismissive of emotional expressions.[19] The problem with reactions of disgust, repugnance, or like feelings is that they are a poor guide to collective action and public policy. Their seemingly subjective character further impedes an appropriate legal response, given the impersonal nature of law. Such feelings could well be related to political and ideological views of the world, but both feelings and views change over time.[20] The HA Consultation Paper attempted to present an ethical discussion in a balanced manner by clearly setting out and addressing commonly articulated or anticipated concerns and fears.[21] Substantive issues have been introduced and counter-arguments presented in order for the discussion to be rounded on the whole. Still, in a meeting with religious group leaders on 13 August 2008, it was observed by some that the ethical discussion in the HA Consultation Paper came across as 'consequentialist' in general orientation. This perception may be attributable to the categorical (and taxonomic) approach that sought to balance current and potential uses against anticipated risks. This was also a critique, by some bioethicists, of Henry Greely's allegedly utilitarian approach.[22] It is questionable if public

---

[18] Maclaren K. Emotional Metamorphoses: The Role of Others in Becoming a Subject, in *Embodiment and Agency*, eds. S Campbell, L Meynell and S Sherwin. University Park, PA: Pennsylvania State University, 2009, pp. 25–45, p. 26. She reasons: "Emotion is not an internal conscious event, but rather the experience of a tension within our reality that puts into question our place in reality. Expressions of emotion, correlatively, are not a matter of indulging in some irrational inner feeling or force, but rather a matter of trying to make sense of our situation, given the resources that we have. The compulsions and primitive behavior that sometimes result when emotions 'escalate' or 'spiral downwards' are to be explained not as merely irrational, but in terms of an increasingly constrained situation, where the emotional person finds herself increasingly stripped of the resources that she assumed she had for making sense of this situation. Other people play an essential role in producing such a constrained situation . . . But at the same time, others can lend us new existential resources for making sense of our situation" (p. 42).

[19] Maclaren indicates that emotional responses are often assumed to be simply irrational ways of configuring reality, and that a person already knows better, or already has access to the 'rational' response. *Ibid.* p. 43.

[20] Jones D. Moral Psychology: The Depths of Disgust, *Nature* **447**, 7146 (14 June 2007): 768–771. Jones writes: " . . . data from psychology and neuroscience should make us think twice about drawing on revulsion as a basis for our personal moral judgements. History seems to bear this out. Women (especially menstruating ones), the mentally and physically disabled, and interracial sex have all been viewed with disgust, and are still viewed as such by some . . . If disgust wasn't a good moral indicator then, why should it be now?" (p. 771).

[21] Fieldnotes, 21 August 2007 (Ho WC).

[22] Baylis F and Robert JS. Part-Human Chimeras: Worrying the Facts, Probing the Ethics, *American Journal of Bioethics* **7**, 5 (2007): 41–45. Baylis and Robert state: " . . . the general utilitarian framework relied upon by Greely and colleagues in their analysis of the ethics of creating

policy could comprehensively capture all ethical concerns.[23] As the HA Consultation Paper seeks to update the BAC's 2002 report on stem cell research and cloning, it is likely that the principles of justness and sustainability will continue to guide the BAC's deliberation.[24]

The application of these principles is perhaps evident in the ethical identification of interests. An issue arose as to whether a discussion of 'Imago Dei' (Image of God) from an expert paper should be incorporated into the HA Consultation Paper. The concern was that this would be too targeted at a particular community (ie. those of the Christian faith).[25] There is no similar concept of 'Imago Dei' in the Islamic faith, and although there are deities with human-animal forms in Buddhism and Taoism, this did not necessarily imply that Buddhists and Taoists would be more receptive to research involving human-animal combinations since the mixing of human and animal features tend to have negative connotations in Chinese culture. This concern did not find ready or expedient solution, but the discussion on 'Imago Dei' was not specifically raised in the HA Consultation Paper. It was nevertheless implicit in the ethical discussion on objecting to the research due to repugnance or the question of 'playing God'.[26] The fact that 'God' has been set out in upper case suggests the Abrahamic conception of a monotheistic deity.

As the intent was to keep discussion in the HA Consultation Paper open-ended, effective regulation was more difficult to present as it could be seen as pre-empting the discussion if specific regulatory approaches were proposed.[27] The significance attributed to regulatory control was nevertheless

human neuron mice . . . is not, in our view, sufficiently rich as to capture the range of ethical concerns. The ethical concerns with this research are not just about weighing putative harms and benefits . . . chimeric research raises deep philosophical questions about what it means to be human and these questions cannot be addressed by appeal to utility maximizing strategies" (p. 44).

[23]JM Elliott (this publication, closing chapter) argues that it is not the business of policy to attempt any reconciliation of individual ethical concerns.

[24]The BAC indicates that its recommendations are intended to lead to results that are 'just' and 'sustainable'. The former favours research with tremendous potential therapeutic benefits to mankind while the latter requires research to have little biological or genetic impact on future generations. See Bioethics Advisory Committee, Singapore, *Ethical, Legal and Social Issues in Human Stem Cell Research, Reproductive and Therapeutic Cloning*, 21 June 2002, p. 35, para 47.

[25]Nuyen AT. Stem Cell Research and Interspecies Fusion: Some Philosophical Issues, 2007, p. 3. Available at http://www.bioethics-Singapore.org.

[26]Bioethics Advisory Committee, Singapore, *Human-Animal Combinations for Biomedical Research: A Consultation Paper*, 8 January 2008, pp. 20–22.

[27]Fieldnotes, 14 November 2007 (Ho WC).

clear.[28] Instead of recommending specific regulatory approaches, regulatory principles were discussed. Paragraph 56 of the HA Consultation Paper illustrates this approach:[29]

> *Most if not all forms of biomedical research involving human subjects pose a threat to the dignity and integrity of human beings at some level. However, such research is not the subject of a comprehensive ban because the risk of serious harm can be mitigated by an effective legal and regulatory regime. In addition, this regime is increasingly supported by a more pervasive ethical infrastructure, within which research is also reviewed by research ethics committees or Institutional Review Boards (IRBs). An example of what such an ethical infrastructure attempts to achieve is encapsulated in the recommendations of the International Society for Stem Cell Research (ISSCR). These recommendations seek to ensure that all human embryonic stem cell research, whether or not human-animal combinations are used, meets certain requirements. They include scientific merit, being directed to the increase of knowledge and potential public benefit, taking place in appropriate facilities with properly trained and supported scientists and staff, and having been peer reviewed.*

This paragraph is also important as a reflection of the view that research involving human-animal combinations should — as a baseline standard — remain governed within the existing ethical framework. Hence, four factors to be taken into account in the creation of human–non-human primate neural tissue chimeras via the implantation of human neural stem cells into an animal are stated essentially as ethical considerations. These factors are:[30]

1. Proportion or ratio of human to animal cells in the animal's brain;
2. Site of integration of the human neural cells;
3. Recipient species; and
4. Brain size of the animal involved.

[28] *Ibid.*
[29] Bioethics Advisory Committee, Singapore, *Human-Animal Combinations for Biomedical Research: A Consultation Paper*, 8 January 2008, p. 27.
[30] *Ibid.* p. 23, para 43.

In addition, the regulatory measures adopted in other jurisdictions were also summarised in the HA Consultation Paper. The Executive Summary of the HA Consultation Paper emphasised the importance of effective regulatory safeguards in the event that research involving human-animal combinations is allowed. It also made clear what would not be permitted, such as allowing human-animal combinations to develop to term or for them to be implanted into a womb. The possibility of setting out legal parameters to enable the measured advancement of science was used against the 'slippery slope' concern.[31]

CURRENT REGULATORY FRAMEWORK IN SINGAPORE

A basic regulatory framework is already in place, in the form of an assemblage of legal provisions under the *Human Cloning and Other Prohibited Practices Act* (Singapore Cloning Act),[32] MOH regulations on assisted reproduction (AR Directives),[33] and ethical requirements set out in the Stem Cell Report. Under the legal-regulatory framework, the following are prohibited:

1. Human reproductive cloning;[34]
2. Developing any human embryo created other than by fertilisation of a human egg by a human sperm for more than 14 days, excluding any period when the development of the embryo is suspended;[35]
3. Trans-species fertilisation for the purpose of reproduction. Where this is done to assess or diagnose sub-fertility, the resultant hybrid must be terminated at the two-cell stage;[36] and

---

[31] *Ibid.* p. 29, paras 60–63.
[32] Singapore Statutes: *Human Cloning and Other Prohibited Practices Act* (Cap. 131B), Revised 2005.
[33] Ministry of Health, Singapore, *Directives for Private Healthcare Institutions Providing Assisted Reproduction Services: Regulation 4 of the Private Hospitals and Medical Clinics Regulations* (Cap. 248, Reg1), Revised 31 March 2006.
[34] Singapore Statutes: *Human Cloning and Other Prohibited Practices Act* (Cap. 131B), Revised 2005, Section 5.
[35] Singapore Statutes: *Human Cloning and Other Prohibited Practices Act* (Cap. 131B), Revised 2005, Section 8.
[36] Ministry of Health, Singapore, *Directives for Private Healthcare Institutions Providing Assisted Reproduction Services: Regulation 4 of the Private Hospitals and Medical Clinics Regulations* (Cap. 248, Reg1), Revised 31 March 2006, para 8.7.

4. Developing a cytoplasmic hybrid embryo beyond 14 days, excluding any period when the development of the embryo is suspended.[37]

The relevance of restricting the development of a hybrid from trans-species fertilisation is questionable in view of possible application of the 14-day or equivalent rule to such hybrids. The trans-species fertilisation (hamster egg penetration) test developed in the 1970s is much less used as new assisted reproductive technologies are now available. Apart from this, there does not appear to be any good reason to justify this sort of experiment. A possible reason why this restriction still persists today might be the difficulty in persuading the public to accept a seemingly more liberal regulatory stance.

Within this regulatory ambit, the Stem Cell Report specifies the categories of stem cell research that are permitted under varying degrees of regulatory purview. These may be summarised as:[38]

1. Research involving the derivation and use of stem cells from adult tissues, subject to the informed consent of the tissue donor;
2. Research involving the derivation and use of stem cells from cadaveric foetal tissues, subject to the informed consent of the tissue donor. The decision to donate the cadaveric foetal tissue must be made independently from the decision to abort;
3. Research on existing embryonic stem cell lines, derived from embryos less than 14 days old;
4. Derivation of stem cells from surplus embryos; and
5. The creation of human embryos specifically for research, which can only be justified where:
    a. There is strong scientific merit in, and potential medical benefit from, such research;
    b. No acceptable alternative exists; and
    c. On a highly selective case-by-case basis, with specific approval from the proposed statutory body.

[37] Possibly under Section 5 of the Singapore Cloning Act, depending on how broadly the term 'human embryo clone' is read.

[38] Bioethics Advisory Committee, Singapore, *Ethical, Legal and Social Issues in Human Stem Cell Research, Reproductive and Therapeutic Cloning*, 21 June 2002, pp. vii and viii.

The requirement that "no acceptable alternative exists" is ambiguous as it would depend on who decides what is acceptable. It may now be given a very restrictive reading by IRBs, particularly in view of induced pluripotent stem cell (iPSC) technology that has recently gained prominence. It is also ambiguous as to whether the three conditions set out in sub-paragraph 5 above are applicable only to embryos created by SCNT or more generally to the creation of embryos for research. The creation of embryos for research by SCNT could be interpreted as being subject to a more stringent requirement than the creation of embryos by other means, such as *in vitro* fertilisation (IVF).

It used to be thought that a main source of pluripotent stem cells would be embryos. And if such pluripotent stem cells were to be patient-specific, then SCNT was considered to be an important technique that would enable the derivation. However, in November 2007, a group of Japanese researchers announced a means by which human pluripotent stem cells could be derived without using embryos.[39] In essence, adult dermal fibroblasts (skin cells) were re-programmed (induced) to function like pluripotent stem cells through the use of viral factors. This represented an important proof of principle that pluripotent stem cells can be generated from somatic cells by the combination of a small number of factors. Such cells were referred to as iPSC.

This development immediately raised a question as to whether SCNT had been rendered obsolete. It was at that time still unclear if the technology could be used to derive human iPSCs from other types of human tissue. Even if this could be done, the level of pluripotence of iPSCs might not be as effective or efficient as stem cells derived through SCNT . In the light of these uncertainties, the BAC did not delay its planned public consultation on human-animal combinations. It provided this explanation in the HA Consultation Paper:[40]

> *Nuclear reprogramming of somatic cell nuclei without the use of SCNT , and thus without requiring human eggs, has recently been reported. Research groups demonstrated that human skin cells can be transformed into cells with properties similar to that of embryonic stem cells through the introduction of specific genes into the*

[39]Takahashi K *et al.* Induction of Pluripotent Stem Cells from Adult Human Fibroblasts by Defined Factors, *Cell* **131**, 5 (2007): 1–12. See also Yu J *et al.* Induced Pluripotent Stem Cell Lines Derived from Human Somatic Cells, *Science* **318**, 5858 (21 Dec 2007): 1917–1920.
[40]Bioethics Advisory Committee, Singapore, *Human-Animal Combinations for Biomedical Research: A Consultation Paper*, 8 January 2008, p. 11, para 8.

> *skin cells. The transformed cells are called induced pluripotent stem cells. This technology could lead to the creation of patient-specific and disease-specific pluripotent stem cells and is a welcome development, although it remains to be seen to what extent it will lead to reduced SCNT research.*

It is also questionable as to whether the hierarchy of sources for human embryonic stem cells[41] remains relevant (ie. a concept that surplus embryos should be used before an embryo may be created for research). One view is that there are insufficient leftover embryos from IVF treatment for research. Most of these embryos are not donated for research, and this problem is not peculiar to Singapore. Thus the statement "As long as there are sufficient and appropriately donated surplus embryos from fertility treatments available for use in research . . . "[42] could be inaccurate in the light of current experience. It is not only a question of numbers. Potential uses of stem cells include allowing researchers to study and understand the processes in developmental biology, to test new drugs, and to generate cells and tissues for therapy. In particular, disease-specific cell lines would enable researchers to understand the disease in question and find better ways to treat it. For instance, if researchers want to study some form of pancreatic abnormality, they would want to be able to create pancreatic cells from someone with a specific disease in order to see in what way their cells do not work. Further down the road, scientists will want to create cells from a particular patient for transplantation without fear of rejection. For these reasons, researchers will want to choose the individuals from whom they can create the cell lines, and this cannot be done by working on some random embryo out of the cell bank even if there is consent for its use. There must be specific ways to create disease-specific cell lines.

CATEGORISATION AND CLASSIFICATION

For the purposes of public consultation, a neutral definition of 'chimeras' and 'hybrids' in the Chinese language was thought to be necessary since the subject matter was already being discussed in that language forum. The expression '嵌合体' is commonly used in scientific and ethical discussions in

---

[41] Bioethics Advisory Committee, Singapore, *Ethical, Legal and Social Issues in Human Stem Cell Research, Reproductive and Therapeutic Cloning,* 21 June 2002, p. vii, reading Recommendations 4 and 5 of the Stem Cell Report together.
[42] *Ibid.* p. 28, para 30.

both China and Taiwan, and is also the expression used in the newspaper commentary noted earlier. In addition, this expression has been applied in a number of English-Chinese dictionaries. In Japanese, Kanji (old-script Chinese) characters have not been used to depict 'chimera' although they are used for the more generic reference to 'cell' (细胞). Instead, the term 'chimera' is set out in katakana as 'キメラ', which suggests the importation of a foreign term (and its corresponding meaning) into the Japanese lexicon. For instance, this term has been used by Japan's Ministry of Education, Culture, Sports, Science and Technology (文部科学省).[43]

Chimeras and hybrids are metaphors that enable the transmission of these concepts from myths and imagination to present-day reality. Even though a chimeric mouse, let alone a cytoplasmic hybrid embryo, does not look anything like Homer's Medusa or the characters in the Chinese classic 西游记, *Journey to the West*, the ontologically creative function of metaphors enable a cognitive equivalence to be drawn between the two metaphoric objects.[44] Hence, in the public imagination, the creation of chimeras or hybrids by scientists is no different from the creation of monsters. Véronique Mottier's observation that metaphors inform and structure thinking as "mini-narratives" acting against a backdrop of tacit knowledge is pertinent.[45] These "mini-narratives" have the capacity to form identities through discursive mechanisms of boundary drawing, boundary maintenance, ordering, and othering. A number of well-known Chinese didactic

---

[43] See, for instance, 文部科学省. 生命倫理及び安全対策に係る留意事項 [Ministry of Education, Culture, Sports, Science and Technology, Japan. Notice on Bioethics and Safety Considerations]. Available at: http://www.mext.go.jp/a_menu/kagaku/chousei/unyo/kobo/13/koubo13/05/0501.htm. See also 文部科学省. 人クローン胚の研究目的の作成利用のあり方に関する検討経緯等について [Ministry of Education, Culture, Sports, Science and Technology, Japan. Background Discussion on the Purposes and Uses of Human Cloned Embryos]. Available at: http://www.mext.go.jp/b_menu/shingi/gijyutu/gijyutu1/006/gijiroku/06060909/001.htm. The reference to chimeric embryos in Japanese (キメラ) appears to be a more generic reference to human-animal combinations (嵌合胚胎) in Chinese.

[44] Relying on the seminal work of Max Black, Terrell Carver and Jernej Pikalo indicate that metaphors are creative-productive function that shape thinking on discourses and contexts. They also act as discursive nodal points. See Carver T and Pikalo J. Editors' Introduction, in *Political Language and Metaphor: Interpreting and Changing the World*, eds. T Carver and J Pikalo. London and New York: Routledge, 2008, pp. 1–12, pp. 3–4. See also Max Black. *Models and Metaphors: Studies in Language and Philosophy*. Ithaca and London: Cornell University Press, 1962.

[45] Mottier V. Metaphors, Mini-narratives and Foucaldian Discourse Theory, in *Political Language and Metaphor: Interpreting and Changing the World*, eds. T Carver and J Pikalo. London and New York: Routledge, 2008, pp. 182–194, pp. 191–192.

folklores with imaginary entities (that include foxes and ghosts) have these multifold, and evidently moralising, functions.[46] In our present situation, the familiar mythical metaphors constitute the identities of the biological constructs. They pose the most direct and immediate challenge as models of some prior and typically unarticulated understanding of the phenomena — perhaps related in some way to the 'yuk' feeling that some have raised in opposition to the creation of human-animal combinations in research. Dvora Yanow defines "models *of* prior conceptualization" as metaphors, or forms of "seeing-as". These metaphors "embody and reflect context-specific prior understanding of their subject matter, drawing — usually implicitly, through tacit knowledge — on metaphoric meaning in its source origins".[47] More insightful is her recognition that metaphors are also "models *for*", in that "they embody seeds for subsequent, future action that follows from the underlying logic of the prior understanding on which they draw".[48] In order to counter the existing metaphors or models *of* chimeras and hybrids, it would be necessary to discredit them. Once discredited, these ideas would then have to be disembedded from the familiar mythical context and reconstituted within a new context that will enable the new to be understood in terms of the old. It may be inferred from the tabular display (Table 2) of human-animal combinations that the new context is grounded in the rationalities of ethics and science combined, and applied toward the maximisation of the common good (in that the end goal is some therapeutic benefit). In contrast, the old context is projected as speculative, irrational, and superstitious.

During the public consultation, the attempt to displace more commonplace notions of chimeras and hybrids was on the whole successful. It subsequently emerged during the public consultation that neither the English language nor Chinese language newspapers deployed the more common term 'chimera', but instead preferred a terminology that captured the essence of the biological construct. The English language papers used the term 'human-animal combinations' that was proposed by the BAC and attempted

[46]Chan LT. *The Discourse on Foxes and Ghosts: Ji Yun and Eighteenth-Century Literati Storytelling.* Hong Kong: Chinese University Press, 1998. See, for instance, pp. 246–247.
[47]Yanow D. Cognition Meets Action: Metaphors as Models of and Models for, in *Political Language and Metaphor: Interpreting and Changing the World,* eds. T Carver and J Pikalo. London and New York: Routledge, 2008, pp. 225–238, p. 227 (emphasis in original).
[48]*Ibid.*

to present the discussion in a relatively objective light. This was similarly the case for the mainstream Chinese language newspaper (联合早报) in Singapore.[49] Two other Chinese language newspapers attempted to sensationalise the subject with controversial cartoons of creatures such as a pig with a human head, but the content of these articles was a reasonable presentation of the issues.[50] News reports from China, Taiwan, and Hong Kong, as well as the news coverage by the main Chinese language newspaper in Singapore, referred to human-animal combinations as '人兽混合体'. There is some ambiguity in relation to the term '嵌合体', which has been defined to mean chimeras, but has also been used to refer to human-animal combinations in general. In relation to the HA Consultation Paper, the Chinese press used the term '嵌合体' to refer to human-animal combinations generally. The two main categories of human-animal combinations considered by the BAC have been translated by the Chinese press as '杂合体' for hybrids and '客迈拉' for chimeras. Cytoplasmic hybrid (or cybrid) embryos, which are a sub-category of hybrids, were translated as '胞质杂配胚胎'. This is consistent with the terminology used in China, notably by the Chinese Medical Doctor Association.[51] It follows that cytoplasmic hybrids are translated as '胞质杂合体'. In some reports, a more generic term '人兽混合胚胎' is used to describe cytoplasmic hybrids as human-animal interspecies embryos. Hence, a function of the HA Consultation Paper was to displace the 'old' and conventional metaphoric notions of chimeras and hybrids with ethically and scientifically 'appropriate' ones.

The approach adopted by the BAC may also be described as an analytical (or displacement) technique that sets out in taxonomies the types of chimeras and hybrids that occur naturally, those that have been created by scientists,

[49]谢燕燕. 生物道德资询委员会征询公众: 你是否接受人兽嵌合体, 联合早报 [Xie YY. Bioethics Advisory Committee Consults the Public: Can You Accept Human-Animal Combinations, *Lianhe Zaobao*], 9 January 2008.
[50]许翔宇. 人兽混合体? 生物道德资询委员会征询公众意见, 联合晚报 [Xu XY. Human-Animal Combinations? Bioethics Advisory Committee Seeks Public Opinion, *Lianhe Wanbao*], 8 January 2008. See also 郭秀芳与李腾宝. 人兽怪研究, 引发各种问题: 带人细胞的肉, 你敢吃吗? 新民晚报 [Xu XF and Li TB. Human-Animal Strange Research, Raises All Kinds of Issues: Meat with Human Cells, Do You Dare Eat?, *Xinmin Wanbao*], 8 January 2008.
[51]中国医师协会, 人兽混合胚胎问世 [Chinese Medical Doctor Association, Birth of Human-Animal Hybrid Embryo], 3 April 2008. Available at: http://www.cmda.net/info/showinfo.asp?id=5520. See also 你能否接受人兽嵌合体? [Can You Accept Human-Animal Combinations?], 1 September 2008. Available at: http://biology.aweb.com.cn/news/2008/1/9/1112193.shtml.

and those that could be created. This passage from the summary of the HA Consultation Paper is illustrative:[52]

> *... a chimera is an imaginary creature, made up of parts from two or more different species, for example a centaur, with the body of a horse and a human head and torso. To Singaporeans, the Merlion is a familiar chimera. However, when scientists talk about human-animal combinations in research, they do not plan the creation of such monsters. In science, a chimera is an animal or a human whose body contains cells or tissues from another animal or human. Any person who has undergone a blood transfusion or any kind of transplant is by definition a chimera, because his or her body would contain cells or tissue from the donor... A well known animal hybrid is the mule, which is the product of crossing a horse and a donkey.*

Being somewhat 'natural', a recipient of blood transfusion and a mule would technically be a 'chimera' and a 'hybrid' respectively, without provoking any feeling of disgust in contemporary society. Categorisation based on 'natural' as opposed to 'unnatural', and further related to current and potential uses, reflects Henry Greely's approach. He similarly attempts to assess how taxonomy of chimeras might illuminate ethical issues that categories create by relying on four dimensions based on biological constituents, relationship between the organisms, how mixing is done, and when it takes place.[53] Greely concludes that the ethical issues depend on the 'humanity', 'naturalness', and proposed uses of the chimeric organism. The approach of the BAC bears some semblance to Greely's and, as we have discussed, was viewed by some to have a more consequentialist character.

---

[52] Bioethics Advisory Committee, Singapore, *Human-Animal Combinations for Biomedical Research: A Consultation Paper,* 8 January 2008, p. 6. A description of the history of the Merlion and its symbolic significance to Singapore is available at: https://app.stb.gov.sg/asp/form/form01.asp.

[53] Greely HT. Defining Chimeras... and Chimeric Concerns, *American Journal of Bioethics* **3**, 3 (2003): 17–20.

REACTION OF THE SINGAPOREAN PUBLIC

Public consultation in Singapore on the subject of human-animal combinations concluded on 10 March 2008. Similar to previous public consultations of the BAC, written submissions were received from scientists, lay members of the public, and institutions. In addition, 58 entries were made on the REACH Discussion Forum[54] by at least 43 individuals.[55] Most comments on REACH are made anonymously (or pseudonymously).

Of the 43 individuals who provided feedback on the REACH Discussion Forum, a majority (18) expressed some support for research using human-animal combinations. However, a large number of individuals (12) did not express any view, whereas the number of individuals who were neutral or did not express a clear view on the issue (7 respondents) is close to the number of individuals opposed to the research (6 respondents). In addition, six comments were received via REACH on the Consultation Paper in general. Of these, five expressed opposition to the research.

The main reasons or conditions for supporting the research were:

1. Therapeutic benefit that may be gained;
2. Confers competitive advantage on Singapore as a leader in biomedical science;
3. Provided that reliable and stringent regulations are in place;
4. Provided that there is a system of governance that is transparent, accountable, and verifiable (by non-scientists, as it may not be realistic to rely on scientists to regulate themselves);
5. Provided that the research is monitored through self-regulation (ie. by other qualified scientists);
6. Provided that animal welfare is assured;
7. Provided that research is meaningful and poses minimal risk;
8. Provided that there is public support;
9. Provided that safety concerns are addressed; and
10. Provided that there is a high degree of caution.

---

[54]REACH (Reaching Everyone for Active Citizenry @ Home) was set up by the Feedback Unit in 2006 to engage and reach out to as many Singaporean and permanent residents as possible to develop and promote an active citizenry through citizen participation and involvement.

[55]Repeated entries on the REACH Discussion Forum were excluded as far as practicable to avoid double-counting.

The main reasons given for opposing the research were:

1. Concern over safety and lack of justification;
2. Concern over animal welfare;
3. Misunderstanding of the science (a respondent indicated that "There may one day be a dog with a human face");
4. Objection to mixing human and animal genetic materials;
5. Those who are in need of the treatment should participate in the research;
6. 'Playing God' and going down a slippery slope;
7. Objection on religious grounds; and
8. Research is considered to be highly unprofitable or not beneficial.

A relatively large number of respondents expressed neither support for nor opposition to the research. They did however provide general comments:

1. Concern over safety, public acceptance, and suitability of the animals used;
2. Concern over the lack of public support;
3. Debate over the relevance of religious views;
4. An indication that those who do not support the research should not enjoy its benefits;
5. Query over the ways in which control may be carried out;
6. An indication that regulatory control should encompass more than the use of force; and
7. The effectiveness of controls was questioned.

On the whole, there was no clear indication of support or opposition for the research. A similar outcome was observed of the British public's reaction to research involving cytoplasmic hybrids. Nevertheless, the outcome of the public consultation was important to the BAC for a number of reasons. First, the absence of strong public reaction was an indication of public receptivity to the research. This may perhaps be attributed to the absence of a dominant group or discourse in the consultation. Second, there was no serious omission in the ethical identification of issues or interests. The strongest reactions for and against the research were for reasons that were within expectation. A level of agreement over ethical rules that can be applied is important for the purposes of legal objectification, as we shall consider. Third, emphasis on sound and effective regulation enabled a practical and measured response to a number of uncertainties and concerns.

RECENT DEVELOPMENTS IN THE UK

In the UK, human-animal combinations in research became a public issue when an expert advisory group recommended that the creation of cytoplasmic hybrid embryos should be prohibited.[56] However, the House of Lords Select Committee on stem cell research did not agree, and suggested that, on the contrary, research using cytoplasmic hybrid embryos might be more acceptable given that human eggs would not be required.[57] The House of Commons Science and Technology Select Committee agreed, and proposed new legislation in 2005 to define the nature of inter-species embryos and to make their creation legal for research purposes, subject to the 14-day rule.[58] In August that year, the Science Media Centre organised a background briefing on chimeras. At this briefing, the possibility of using cytoplasmic hybrid embryos in research was discussed, and concerns over regulatory loopholes were also raised. This was identified as the first step in what became a concerted attempt to keep the public and politicians, via the media, fully informed.[59] By December 2006, however, a government white paper proposed prohibiting the creation of hybrid and chimeric embryos.[60] This development followed applications by researchers from King's College London and Newcastle University for permission to create cytoplasmic hybrid embryos for research. The possible imposition of legal limitations on embryonic stem cell research prompted the HFEA to conduct a full consultation to gauge public opinion,[61] and the pooling together of efforts by the Academy of Medical Sciences, the Wellcome Trust, the Royal Society, the Medical Research Council, the Association of Medical Research Charities, and many individual

[56] Department of Health, UK, *Stem Cell Research: Medical Progress with Responsibility*, June 2000. See Recommendation 6, p. 47.
[57] House of Lords Select Committee, UK, *Stem Cell Research*, 2002. See para 8.18.
[58] House of Commons Science and Technology Select Committee, *Reproductive Technologies and the Law*, 2005, para 66.
[59] Watts G, ed. *Hype Hope and Hybrids: Science, Policy and Media Perspectives of the Human Fertilisation and Embryology Bill*, 2009. UK: The Academy of Medical Sciences, Medical Research Council, Science Media Centre and Wellcome Trust, p. 11.
[60] Department of Health, UK, *Review of the Human Fertilisation and Embryology Act: Proposals for Revised Legislation (Including Establishment of the Regulatory Authority for Tissue and Embryos)*, December 2006.
[61] Human Fertilisation and Embryology Authority, UK, *Hybrids and Chimeras: A Report on the Findings of the Consultation*, October 2007.

patient charities to clarify the basis and aims of such research, and the issues raised. A consultation paper entitled 'Hybrids and Chimeras' was issued by the HFEA to solicit feedback over a period of three months, from 26 April to 20 July 2007.[62] Those who responded to the poll or to public consultation had strong views one way or the other. The full consultation conducted by the HFEA revealed that while up to 67% of the respondents initially opposed creating hybrids in general, opposition fell to 30% and support rose to 50% when it was explained that the research could help scientists understand diseases such as Parkinson's and Alzheimer's disease.[63] From the various polls and public consultations conducted, it was observed that on the whole there was not a clear majority of the British public either opposed to or supporting the research.[64] However, the polls did suggest majority public support for the creation of hybrid embryos when the likelihood of treatments for named diseases was indicated. Where there were no claims of important potential medical advances (such as in the creation of 'true' hybrids), a majority was opposed.

It was during this time that the AMS published a report on the subject of inter-species embryos,[65] to clarify the terminology and the scientific issues in the debate. The main objectives of the AMS were:[66]

- *To propose definitions of embryos combining human and non-human material, and identify relevant research protocols.*
- *To identify and agree key opportunities for research using such embryos, and cells derived from them, together with an assessment of how these opportunities are balanced by safety and ethical concerns.*
- *To provide recommendations where appropriate.*

---

[62] Human Fertilisation and Embryology Authority, UK, Press Statement, 26 April 2007.

[63] Jones DA. What Does the British Public Think About Human-Animal Hybrid Embryos? *Journal of Medical Ethics* **35**, 3 (2009): 168–170, p. 169.

[64] *Ibid.*

[65] The Academy of Medical Sciences, UK, *Inter-specieInter-species Embryos: A Report by the Academy of Medical Sciences*, June 2007. The Academy of Medical Sciences was formed in 1998 to develop and support UK medical science and individual biomedical researchers into the future. As an independent academy of experts, it acts as a champion for exploration of knowledge, a guardian of intellectual rigour and excellence, and an advisor on national public policy issues. It is housed in the impressive No. 10 Carlton Terrace of the British Academy. The area itself is home to many other learned societies and academic organisations including the Institute of Materials, Minerals and Mining, the Royal College of Pathologists, the Royal Academy of Engineers, and the Royal Society.

[66] *Ibid.* p. 5.

The report was put together relatively quickly in view of the political developments on the subject.[67] Consultations were conducted with regulatory authorities like the HFEA. Following the publication of the report, a briefing for Ministers of Parliament was jointly undertaken by the AMS, the Medical Research Council, the Royal Society, and the Wellcome Trust at the House of Commons on 6 May 2008.

The substantive report begins with an explanation of the value of human-animal combinations in the context of human embryonic stem cell research (hESCR) and SCNT . It emphasises that this research is directed at learning how to control stem cell differentiation and development, understanding nuclear reprogramming, and generating specialised tissues in culture or animal models.[68] Human-animal combinations could help overcome the shortage of human oocytes for such research.[69]

Strong arguments are presented for extending the current legislation on human embryo research to inter-species embryos. This was timely in view of the fact that the Human Fertilisation and Embryology (HFE) Act of 1990 was being reviewed in the UK Parliament. Essentially, the position advanced by the AMS is for the creation and use of inter-species embryos, especially cytoplasmic hybrid embryos, to be permissible but subject to legally prescribed limits of:

1. Licensing from the HFEA;
2. Prohibiting implantation into a woman or animal; and
3. Prohibiting the development of such embryos beyond 14 days *in vitro*.

Under such a framework of regulatory control, the AMS indicates that there would be no substantive ethical or moral reasons not to proceed with the research on cytoplasmic hybrid, human transgenic, or human chimeric embryos.[70]

It further indicates that *in vitro* laboratory research involving cytoplasmic hybrids or other inter-species embryos would not raise any significant

[67] Interview with Dr Helen Munn, Executive Director (then Director, Medical Science Policy), Academy of Medical Sciences, 1 April 2008 (Ho WC).
[68] The Academy of Medical Sciences, UK, *Inter-species embryos: A report by the Academy of Medical Sciences*, June 2007, pp. 7–14.
[69] *Ibid.* pp. 16–17.
[70] *Ibid.* pp. 30–31.

safety risks over and above regular cell culture, provided that good laboratory practice is rigorously followed. However, if cell lines derived from such embryos should ever be contemplated for therapeutic use, it would be prudent to scan the mitochondria and cytoplasmic RNA of the species to be used for pathogens.[71]

The governance structure proposed by the AMS to oversee the research is based on the recommendations of the ISSCR.[72] Three categories of review were proposed:

1. The first category relates to experiments that are permissible after review by existing local committees. These experiments include research with pre-existing human embryonic stem cell lines that are confined to cell culture or involve routine and standard research practice;
2. The second category relates to research that is permissible only after additional and comprehensive review by a specialised body. Such research includes mixing human totipotent or pluripotent stem cells with pre-implantation human embryos, and research that generates chimeric animals using human cells at any stage of post-fertilisation, foetal, or post-natal development; and
3. The third category sets out research that is not permissible at the current time. Such research includes implantation of human totipotent or pluripotent cells, or their constituent product, into a human or non-human primate uterus, and research in which animal chimeras incorporating human cells with the potential to form gametes are bred together.

Like the ISSCR, the AMS would not prohibit the introduction of human pluripotent stem cells into a non-human primate or human embryo if there is sufficient justification for the research. It similarly sets out caveats relating to the proportion of human stem cells transferred, the likely integration into critical tissues such as the germline and central nervous system, and the transfer of human cells into non-human primates. If transgenic and chimeric animals contain significant amounts of human genetic material, an appropriate conceptual and regulatory framework will have to be considered.

[71] *Ibid.* p. 38.
[72] *Ibid.* p. 35.

Currently, the creation of transgenic animals (including those incorporating human DNA) in the UK is regulated by the Home Office under the *Animal (Scientific Procedures) Act 1986*. An inter-species embryo would not fall under the remit of the Act as it only applies to an animal from the mid-point of gestation or incubation period for the relevant species.[73] An early animal embryo with a significant amount of human material may therefore fall outside the regulatory framework altogether.[74] Where the introduction of regulatory control is intended, the AMS cautions that rigid categorisation in primary legislation would not be an appropriate regulatory approach, preferring instead the regulatory framework to operate on a case-by-case basis. It further indicated the need to establish proper channels of communication and consultation between those bodies regulating human embryos, human stem cells, and animal research.[75] The report was undertaken as a step in this direction.[76]

The HFE Amendment Bill was introduced in the House of Lords on 8 November 2007. The intent was to update (but otherwise retaining) the "current model of regulation and the basic foundations of the existing law contained in the Human Fertilisation and Embryology Act 1990" in order to "help maintain the UK's position as a world leader in reproductive technologies and research".[77] One of the main elements of the Bill is to increase the scope of legitimate embryo research activities, including 'inter-species embryos'. The Bill initially followed the terminology of the AMS report in its reference to 'inter-species embryos', but this was replaced by 'human admixed embryos'. This change was introduced so as to indicate that a cytoplasmic hybrid embryo is effectively 'human'. This classification is important because the HFEA does not regulate non-human embryo research. And as noted in the AMS report, *in vitro* early animal embryo research does not appear to fall within any regulatory purview. The Bill's provisions for inter-specific hybrids were not expected to be especially controversial in the UK, although there was a vocal minority in strong opposition. On the whole, there

[73] *Ibid.* p. 33.
[74] *Ibid.* p. 40.
[75] *Ibid.* pp. 38–39.
[76] *Ibid.* p. 40. In the concluding section of the report, it states: "As a first step, it is essential that proper channels of communication and consultation are established between those bodies regulating human embryos, human stem cells and animal research."
[77] Shepherd E. *Human Fertilisation and Embryology Bill [HL], HL Bill 6, 2007–08*. London: House of Lords Library Notes, 14 November 2007, p. 1.

was relatively strong support for the HFE Amendment Bill. In fact, many of the legislative amendments proposed in the Bill related to more controversial sociological issues, such as marital status and the welfare and rights of children. The provisions that related to human-animal combinations were regarded as technical manoeuvres to ensure that these constructs would fall within the regulatory purview of the HFEA. Hence, the revision of the term 'inter-species embryos' to 'human admixed embryos' is intended to refer to human embryos with some non-human genetic material, as opposed to animal embryos with human material. As such, the Bill was specifically concerned with human embryos and so would not include transgenic animals.

As animal chimeras are not included within the purview of the Bill, future policy and regulatory attention may be required, especially in relation to the introduction of human cells (such as neural cells) into an animal (such as a monkey). Such chimeras, which have nothing to do with human embryology, are more alive in the public imagination, as shown in the press coverage of the BAC's public consultation on human-animal combinations. Experiments involving animals are subject to tight bureaucratic control in the UK, mainly relating to animal welfare, housing, husbandry, and related matters. Research involving human-animal combinations after mid-gestation would fall within the regulatory purview of the authority that controls the use of research animals, only if they were regarded as essentially animals with added human genetic material. And since animal chimeric embryos are not human embryos, they also fall outside the regulatory purview of the HFEA. There might be a need to examine the co-ordination of these two different regulatory regimes to address concerns in animal chimera research. In addition, it would be important not to confuse research involving human-animal combinations at an embryonic level with research using animal chimeras. The UK will have to deliberate on these issues in the near future, as there have already been, for instance, reports of significant amounts of human neural cells being introduced into non-human primates. Such research is contentious in the eyes of the public. This regulatory gap has been identified in the AMS report as a matter for future consideration. Reflecting the concerns of researchers in general, the AMS report indicates that care should be taken not to interfere with certain well-established use of human-animal combinations such as transgenic animals. Careful working through is required to determine where the problems lie, what the regulations should be, who the controller should be, and related issues.

In January 2008, the HFEA approved two applications to create human-animal cytoplasmic hybrids for research.[78] The research teams from King's College London and Newcastle University were each granted one-year research licences. The team from King's College London planned to derive disease-specific stem cell lines from cytoplasmic hybrids to study neuro-degenerative diseases such as Alzheimer's disease, Parkinson's disease, and spinal muscular atrophy. Similarly, the team from Newcastle University hoped to study the process of nuclear reprogramming. Preliminary data was subsequently presented by this team at a lecture at the Knesset in Israel in March 2008, showing that they had created human-animal hybrid embryos by introducing human somatic nuclear material into enucleated cow eggs.[79] Later in June 2008, they reported at the 'BIO' biotechnology conference in San Diego that human-animal admixed embryos are easy to produce on a large scale, and that they had already created about 270 such embryos.[80] They attributed the success to the large number (200) of cow eggs available per day compared to the number (10) of human eggs available per month. However, the embryos stopped growing at the 32-cell stage, and the team has yet to derive stem cells from such embryos. In July 2008, a third team, from the University of Warwick, received a 12-month licence from the HFEA to create human-pig embryos for the study of heart diseases.[81] Skin cells from patients with a mutation for a heart disease (cardiomyopathy) would be placed into pig eggs which had had their nuclear DNA removed. The resulting embryos would be sources from which embryonic stem cells could be derived, so as to study the molecular mechanisms associated with the heart disease.

In January 2009, it was reported[82] that a lack of funding was threatening progress in research involving cytoplasmic hybrids. The teams from King's

[78] Human Fertilisation and Embryology Authority, UK, *Research Licence Committee Meeting*, 28 November 2007.

[79] Newcastle University, UK, *Press Release*: *Hybrid Embryos Statement*, 1 April 2008.

[80] Maden C. Human/Animal Hybrid Embryos are 'Easy' to Make, *BioNews*, 23 June 2008. Available at: http://www.bionews.org.uk/page_13422.asp.

[81] Highfield R. Human-Pig Hybrid Embryos Given Go Ahead, *Telegraph*, 1 July 2008. Available at: http://www.telegraph.co.uk/science/science-news/3345954/Human-pig-hybrid-embryos- given-go-ahead.html.

[82] Wardrop M. Human-Animal Clone Research Halted Amid Funding Drought: Britain's Research into the Creation of Human-Animal Clones Has Ground to a Halt Due to a Lack of Funding, *Telegraph*, 13 January 2009. Available at: http://www.telegraph.co.uk/science/science-news/4225636/Human-animal-clone-research-halted-amid-funding-drought.html. See also Shaikh N. Lack of Funding Threatens Progress in Admixed Embryo Research, *BioNews*, 19 January 2009. Available at: http://www.bionews.org.uk/page_13636.asp.

College London and Newcastle University have been declined funding for their research by the Medical Research Council (MRC) and the Biotechnology and Biological Sciences Research Council (BBSRC). Despite having created 278 admixed embryos, the team from Newcastle University lacked the funds to buy the necessary equipment in order to proceed. The actual reason for the rejection of funding for the research is unclear. It is the policy of the MRC and BBSRC not to comment on individual applications and the two teams were told that their research had to compete with other projects. Having a HFEA licence to conduct research does not automatically entitle researchers to funding. By October 2009, it was reported that all research involving the creation of cybrids had been refused funding in the UK.[83] The key researcher from the team at Warwick University (one of the three teams licensed by the HFEA to carry out such research) was reported to have moved to Australia. The other licence holder from King's College London was reported to have left university for industry, while the licence holder from Newcastle University has moved to Spain.

Nevertheless, the concerted effort of scientific bodies and medical charities in the UK culminated in the successful passage of comprehensive legislation on the subject through Parliament. The HFE Act 2008 was passed in October 2008 and received the royal assent in November 2008.[84] It supersedes the 1990 version of the legislation. Of relevance to the BAC is Part 1 of the legislation, which ensures that all research on human embryos that occur outside the body, and research on human admixed embryos, where animal DNA is not predominant, are subject to strict regulation. The Act, which is administered by the HFEA, will be implemented in stages, with the legislative provisions that relate to research on embryos being effective from October 2009.

In the Act, the term 'embryo' means a live human embryo and includes "an egg that is in the process of fertilisation or is undergoing any other

[83] Conner S. Vital Embryo Research Driven Out of Britain: Scientists Abandon Plan to Develop Stem Cells After Funding Dries Up, *The Independent*, 5 October 2009. Available at: http://www.independent.co.uk/news/science/vital-embryo-research-driven-out-of-britain-1797821.html. See also Blackburn-Starza A. British Scientists Go Abroad Over Lack of Funding for Human Admixed Embryo Research, *BioNews*, 12 October 2009. Available at: http://www.bionews.org.uk/page.asp?obj_id=49593.

[84] Human Fertilisation and Embryology Act, UK, 2008. See also Explanatory Notes to the legislation.

process capable of resulting in an embryo",[85] and a 'human admixed' embryo is:

> (a) *an embryo created by replacing the nucleus of an animal egg or of an animal cell, or two animal pronuclei, with —*
>
> > (i) *two human pronuclei,*
> > (ii) *one nucleus of a human gamete or of any other human cell, or*
> > (iii) *one human gamete or other human cell,*
>
> (b) *any other embryo created by using —*
>
> > (i) *human gametes and animal gametes, or*
> > (ii) *one human pronucleus and one animal pronucleus,*
>
> (c) *a human embryo that has been altered by the introduction of any sequence of nuclear or mitochondrial DNA of an animal into one or more cells of the embryo,*
> (d) *a human embryo that has been altered by the introduction of one or more animal cells, or*
> (e) *any embryo not falling within paragraphs (a) to (d) which contains both nuclear or mitochondrial DNA of a human and nuclear or mitochondrial DNA of an animal ("animal DNA") but in which the animal DNA is not predominant.*[86]

Reference to human cells or animal cells means either cells from a human or human embryo, or cells from an animal or animal embryo, as the case may be. Thus, human admixed embryos, which are created from a combination of human and animal genetic material, would include:

1. Human cytoplasmic hybrid embryos;
2. True human-animal hybrid embryos;
3. Transgenic human embryos;
4. Human-animal chimeric embryos; and
5. Any other embryo that has human and animal nuclear or mitochondrial DNA, but in which the animal DNA is not predominant.

---

[85] *Ibid.* Section 1(2).
[86] *Ibid.* Section 4A (6).

This Act does not cover the creation or use of animal chimeric foetuses or animals, or (as indicated above) the creation or use of human admixed embryos in which animal DNA is predominant.

The creation and use of human admixed embryos covered under the legislation is allowed under licence if they are purely for laboratory research, and such research should not extend beyond 14 days of embryonic development. These embryos are to be destroyed after the 14-day limit is reached.[87] The Act also stipulates that a licence cannot authorise placing a human admixed embryo in a woman[88] or in an animal.[89]

TRANSNATIONAL NORM-MAKING AND THE ISSCR

The guidelines of the ISSCR were considered and cited by both the BAC and the AMS. The ISSCR was formed in 2002 as an international professional organisation of stem cell scientists. It meets annually, mainly in North America, and has co-organised regional meetings. In 2007, it launched the journal *Cell Stem Cell* as a high-end forum for stem cell biology. A year later, experts in law and public policy were involved in its 'Global Forum' series, and a Global Advisory Council was formed to help the ISSCR set a strategy for advancing stem cell research and public education.[90] It was observed that since its founding, the ISSCR

> *remains the only international membership society that embraces researchers, clinicians, ethicists and policymakers, industry representatives, civic leaders, and patient advocates. This broad constituency both motivates and enables the ISSCR to present a united voice for the continued global support of stem cell research and the exchange of scientific information.*[91]

Stem cell researchers in Singapore have been involved in the scientific meetings of the ISSCR. On a number of occasions, a BAC member has also

---

[87] *Ibid.* Sections 4A (2) and 4A (3).
[88] *Ibid.* Section 4A (1) (a).
[89] *Ibid.* Section 4A (4).
[90] Daley GQ, Rooke HM and Witty N. Global Forum Discusses Stem Cell Research Strategy, *Cell Stem Cell* **2**, 5 (2008): 435–436.
[91] Witty N. Strategic Planning: Progress and Potential, *Cell Stem Cell* **1**, 4 (2007): 383–386, p. 383.

been involved in the ethics discussions of the ISSCR. It is in part through such engagement that the views of the ISSCR are important in the deliberations of the BAC.

Three documents provide guidance for the conduct of all hESCR, including research involving cytoplasmic hybrids and chimeras. The first set of guidelines published in December 2006 set out the general regulatory framework for hESCR.[92] They exclude research on animal stem cells, human stem cells that are not known to possess totipotent or pluripotent potential, or the transfer of animal cells into humans at any stage of development. As such, the focus of the ISSCR is on experimental rather than therapeutic application of human pluripotent cells. Similar to the Embryonic Stem Cell Research Oversight (ESCRO) committee of the National Academy of Sciences (NAS),[93] the ISSCR proposes a separate review mechanism for stem cell research, but in the form of a process referred to as Stem Cell Research Oversight (or SCRO).[94] The three categories of stem cell research for experiments that are permissible after standard review, permissible after additional and comprehensive review, and impermissible at this time reflect the ethical framework of the NAS:[95]

1. In the first category, experiments that are permissible after review by existing local ethics committees include research with pre-existing hESC lines that are confined to cell culture, or involve routine and standard research practice;
2. Experiments that fall into the second category are permissible only after additional and comprehensive review by a specialised body, as they require greater levels of scientific justification, as well as more extensive consideration of social and ethical aspects of the research and the reasons for not pursuing alternative methods (if any). Examples of such experiments include mixing human totipotent cells or pluripotent stem

---

[92] International Society for Stem Cell Research, *Guidelines for the Conduct of Human Embryonic Stem Cell Research*, 2006.

[93] National Academy of Sciences, USA, *Guidelines for Human Embryonic Stem Cell Research*, 2005 (amended 2007, 2008, and 2010), p. 217, Section 2.

[94] International Society for Stem Cell Research, *Guidelines for the Conduct of Human Embryonic Stem Cell Research*, 2006, p. 5, para 8 (Recommendations for Oversight).

[95] *Ibid.* pp. 6–7, para 10 (Categories of research). These categories may be compared with those set out in Section 1.3 of the NAS's guidelines on stem cell research. See National Academy of Sciences, USA, *Guidelines for Human Embryonic Stem Cell Research*, 2005 (amended 2007, 2008, and 2010), pp. 14–16.

cells with pre-implantation human embryos (although they are not to be developed *in vitro* for more than 14 days), and research that generates chimeric animals using human cells, including (but not limited to) introducing totipotent or pluripotent human stem cells into non-human animals at any stage of post-fertilisation, foetal, or post-natal development; and

3. As with the NAS, the ISSCR specifies research that is not permissible at the current time in the third category. Such research does not detract from those that the NAS has identified as currently impermissible in that no research should be conducted on an embryo beyond 14 days of development or upon the appearance of the primitive streak, whichever is earlier, no implantation of any product of human stem cell research into a human or primate uterus, and human-animal chimeras should not be allowed to breed with each other. However, the ISSCR Guidelines differ from the NAS Guidelines in that they do not recommend the prohibition of the incorporation of hES cells into non-human primate embryos or of any embryonic stem cell into human embryos. It does specify caveats relating to the proportion of human stem cells transferred, the likely integration into critical tissues such as the germline and central nervous system, and the transfer of human cells into non-human primates.

A year later, the Ethics and Public Policy Committee of the ISSCR published a report that specifically addressed the subject of animal chimeras. It emphasises the importance of animal chimeras as research tools and to avoid "unwanted stem cell exceptionalism".[96] It recommends that human embryos with animal stem cells introduced should not be allowed to develop beyond 14 days or until the formation of the primitive streak and that no products of research involving transferred human cells should be implanted into a human or non-human primate uterus.[97] Thus the moral status of the research embryo up to 14 days of development is not affected by the possible degree of chimerism. The ISSCR is not concerned with the governance of animal usage in experiments, but it specifies that general animal welfare principles continue to apply to the creation of animal chimeras. Reflecting a general concern of developing human sentience in animals, it recommends that research on non-sentient constructs should be employed before research on sentient animals,

[96] International Society for Stem Cell Research. Ethical Standards for Human-to-Animal Chimera Experiments in Stem Cell Research, *Cell Stem Cell* **1**, 4 (2007): 159–163, p. 161.
[97] *Ibid.*

to the extent practicable. Where the research involves creating, gestating, and raising an animal with significantly increased potential for sentience, assessment should be based on a reasonable extension of the standards recognised for research with animal models.[98] Additional data collection and monitoring should be commensurate with the anticipated characteristics of the chimeric animal in the context of the proposed research. Creation of a teratoma to confirm stem cell characteristics should require no additional or exceptional monitoring beyond the application of animal welfare principles, whereas a significant contribution to the central nervous system by hES cells or their direct derivatives will require additional considerations.[99]

Monitoring and data collection should be based upon a sound assessment of the developmental trajectories that are likely to be affected, and take into account the epigenetic context of regulation in which the mixed genes or cells are going to be deployed. Data collection should be linked to known functional links, and links are to be evaluated in a scientifically legitimate manner. The ISSCR observes that no single test, such as the percentage presence of human-derived cells in the brain, should be necessarily required, unless its functional link to pertinent physical or mental qualities is either demonstrated or is consistent with scientific knowledge or scientifically reasonable inferences concerning whether, in the context of other data, it will be a valid predictor of sentience.[100] Chimera neuroscientific research involving human stem cells or their direct derivatives, in hypothetically approximating some aspects of human functioning, may thus demand accepted or new specialised cognitive and other assessments of the sort conducted in neuroscientific research.

The ISSCR recognises that there may also be some degree of manoeuvrability in the way that research can be designed around ethical challenges. Hence, it recommends that investigators and institutions consider requiring the use of pilot studies to produce initial data on chimeric animals subject to experimental interventions, employ ongoing monitoring of deviations from normal behaviours, and prescribe reporting to pertinent animal welfare committees. Investigators and institutions should also make appropriate adjustments to research protocols to take into account new data or unanticipated responses

---

[98] *Ibid.* pp. 161–162.
[99] *Ibid.* p. 162.
[100] *Ibid.*

from animal subjects relating to the ethical permissibility of the animal's participation in the study, including novel signals demonstrating deterioration or enhancement of an animal's condition and other factors pertinent to withdrawing the animal from the study. Regular reassessment of animal welfare during the course of experimentation is strongly encouraged. Research with the known, intended, or well-grounded significant potential to create humanised cognition, awareness, or other mental attributes, while not absolutely prohibited, should be subject to close scrutiny, taking the most careful steps to collect data pertinent to the ethical protection of animal subjects, and an extraordinary degree of justification.[101]

More recently, a report was published to consider if the creation of cytoplasmic hybrids is scientifically justified. This scientific and (to a lesser degree) ethical evaluation was undertaken in the light of some recent developments, notably those in the UK.[102] Contextualising human inter-species SCNT (referred to as 'iSCNT ', where the technique of SCNT is applied to the creation of cytoplasmic hybrids, in contrast to the more conventional hSCNT in therapeutic cloning) alongside the iPSC technique in the broader technology of nuclear reprogramming research, the ISSCR recognises that the latter is a much simpler way of deriving cells with pluripotent qualities. However, there are important differences between the two techniques. First, the questions that these techniques seek to answer in basic biology are different. The epigenetic reprogramming entailed in iPSC is essentially gene-based whereas SCNT is concerned with the oocyte-driven reprogramming of human somatic cell nuclei. Second, SCNT provides important information on human embryogenesis, from which physiological mechanisms and pathological aberrations can be better understood. It is as yet unclear if a viable embryo can be produced from gametes or embryonic stem cells derived from iPSC. Only if such an embryo can be produced from iPSC will there be a clearer basis for re-evaluating the relevance of SCNT . Third, not enough is known about iPSC to "confidently abandon all other research pathways aimed at obtaining pluripotent cells".[103] The ISSCR concludes that rather than rendering SCNT

---

[101] *Ibid.*

[102] International Society for Stem Cell Research. Ethics Report on Interspecies Somatic Cell Nuclear Transfer Research, *Cell Stem Cell* **5**, 1 (2009): 27–30, p. 27.

[103] *Ibid.* p. 28.

obsolete or redundant, the iPSC technique is complementary. It further agrees with the proposal of the UK AMS that:[104]

> *... most scientists agree that the only way to settle the debate about whether these [cytoplasmic hybrid] embryos will produce viable and useful stem cells is to continue conducting the research, perhaps by using different animal eggs and timing nuclear transfer so that genome activations are synchronized... Of course, one could ask how much in iSCNT the admittedly artificial combination of reagents could illuminate the genuine potential of human oocyte-mediated reprogramming. This is a fundamental concern, but one that can only be addressed by performing a thorough comparison between pluripotent cell lines derived by hSCNT versus iSCNT .*

As for the ethical aspects of the research, the ISSCR did not view these to be significantly different from those relating to animal chimeras, even if the latter involved a mixture of genetically different cells rather than a mixture of genetic material within a cell. It did recognise that continued emphasis on the importance of SCNT may result in the exploitation of women for eggs. However, the research should not be precluded only on the basis of this risk if the development of stem cell-based therapies is of social or humanitarian importance. Its reasoning here is instructive:[105]

> *While we agree that potential exploitation of oocyte donors is a serious issue, we maintain that the best course would be to deal with the threat of exploitation squarely by scrutinizing and eliminating recruitment practices that unduly induce women to participate in research, rather than through an unproven and unnecessarily stringent strategy of banning all iSCNT research.*

CHIMERAS AND HYBRIDS AS REGULATORY OBJECTS

In her contribution to the legal portion of the European Union's CHIMBRIDS project, Elisabeth Rynning observes that a difficulty in regulating chimeras and hybrids is the lack of knowledge regarding the

---

[104] *Ibid.*
[105] *Ibid.* p. 29.

potential future value and the risks involved. Consequently, "the necessary balancing of interests to a certain extent will amount to a balancing of two unknowns, making it virtually impossible to reach any well-founded conclusions on proportionality".[106] In managing this uncertainty, categorisation has the effect of lowering the informational deficiency by streamlining a social phenomenon on the basis of certain desired information qualities. Social and cognitive psychologists tell us that this is how we — as human beings — generally store and process information.[107] The generation of categories through rationalisation and systematisation in legal and regulatory processes does not detract significantly from our basic cognitive orientation. Reductionism in categorisation is evident in what Bruno Latour describes as involution in terms of a model of qualification as opposed to fictionalisation.[108] Latour argues that

> *Whereas in science everything is done to ensure that the impact of new information upon a body of established knowledge is as devastating as possible, in law things are arranged in such a way as to ensure that the particular facts are just the external occasion for a change which alters only the law itself, and not the particular facts, about which one can learn nothing further, beyond the name of the claimant.*[109]

The tabularised survey of a range of human-animal combinations based on certain scientific, ethical, and social criteria (in Table 2) is an attempt at acquiring an understanding of the subject at hand for the purposes of steering institutional action. With understanding, it becomes possible to appreciate the regulatory risk[110] entailed as knowledge enables control. Hence, categorisation through typifying human-animal combinations is also an exercise in the enumeration of risks entailed. Objectification was also thereby

[106] Rynning E. Legal Tools and Strategies for the Regulation of Chimbrids, in *CHIMBRIDS — Chimeras and Hybrids in Comparative European and International Research*, eds. J Taupitz and M Weschka. Heidelberg: Springer, 2009, pp. 79–87, p. 85.

[107] Howard JA. Social Psychology of Identities, *Annual Review of Sociology* **26** (2000): 367–393.

[108] Latour B. Scientific Objects and Legal Objectivity, in *Law, Anthropology, and the Constitution of the Social: Making Persons and Things*, eds. A Pottage and M Mundy. Cambridge: Cambridge University Press, 2004, pp. 73–114, p. 104.

[109] *Ibid.* p. 105.

[110] Regulatory risk relates to the risk of not achieving the regulatory objectives or the risk of generating other regulatory issues that are unanticipated when regulation was introduced. See Sunstein CR. Health-Health Tradeoffs, in *Risk and Reason*. Cambridge: Cambridge University Press, 2002, pp. 133–152.

achieved through the substitution of scientific, ethical, and social content in a categorical manner for mythical ones in the metaphors of 'chimeras' and 'hybrids'. As earlier considered, the displacement technique may give description to the BAC's approach in the HA Consultation Paper. Models *for* 'chimeras' and 'hybrids' were constructed to replace models *of* these constructs. The elimination of categories that followed, from the many combinations down to human-to-animal chimeras and cytoplasmic hybrid embryos, effectively reduces the uncertainty in and regulatory risks of the subject matter, facilitating control as well as action.

ARRESTING THE SLIDE DOWNWARDS

Philosopher and BAC member Professor Anh Tuan Nuyen commenced his segment of the public lecture on 16 August 2008 with a travesty of the 'slippery slope' argument.[111] He narrated a scene of two men chatting in a pub. One man told the other that he intended to emigrate. When asked for the reason, he said that homosexuality was initially illegal in the country. But the government had recently decided to repeal the prohibitive legislation and might even decide to allow same-sex marriage. He surmised that given these developments, he should leave the country before he was forced to be gay in the not too distant future.[112] From this vantage point, the flaw in reasoning is apparent. Yet, the metaphorical argument of a 'slippery slope' has been raised in each and every public consultation conducted by the BAC. Feedback from the public consultation also indicates the persuasiveness of this argument in the mind of the public. And this phenomenon is by no means confined to Singapore. Dag Stenvoll provides illustrations of the extensive deployment of this argument in the Norwegian parliamentary debates on sexuality, abortion,

[111]The public forum entitled 'Mixing human and animal tissues: Is such research ethical?' comprised three speakers: a stem cell scientist (Professor Lawrence Stanton), a philosopher (Professor Anh Tuan Nuyen), and a medical ethicist (Professor Bernard Lo). In a newspaper article that publicised this event, it was observed that "Opponents fear that tinkering with the human genome puts science on the slippery slope to creating animal-human hybrids". The BAC's position was then indicated: "But unlike Hollywood science, researchers are not looking to create half-man half-animal monstrosities . . . " See Tan T. Forum on Ethics of Mixing Human, Animal Genes, *The Straits Times*, 14 August 2008.

[112]Certain homosexual conduct are technically illegal in Singapore and could constitute a criminal offence. At around the time of the public lecture, there was ongoing debate as to whether this Edwardian legacy should be repealed. In order to pacify a vocal conservative minority, the government has adopted a pragmatic position of retaining the legislation, but providing an assurance that this legislation will not be enforced.

and new reproductive technologies from the 1950s through the 1990s. His analysis of the 'slippery slope' mechanics is instructive:[113]

> *The slippery slope image works metaphorically in at least two ways. First, it sets up the physical world of solid objects as an analogy to political matters, implying that politics is like the physical world: if you 'move' something in the world of politics, like making or changing a particular law or policy, other things will inevitably follow... Second, the slippery slope does in itself entail a particular image of movement: from a good or relatively good place to a relatively worse or natural world of determinism and laws of physics. It imposes a kind of unidirectional, unstoppable movement which, when used metaphorically about politics, binds phenomena together in a specific way... The metaphorical expression of a slippery slope also involves a dual displacement of focus: regarding time perspective, from the present to the future, and regarding problematisation, from the instant to the danger case.*

When applied in ethical debates, the determinism in Stenvoll's 'politics is physics' is no less forceful as 'ethics is physics'. It became a subject for critique in the editorial of an issue of the *Journal of Medical Ethics*.[114] Whether a scientific possibility is seen as a slope going down towards some detriment or an escalator going up towards some benefit is ultimately a prediction. This misleading metaphoric stealth of hand led to the BAC's decision to avoid using the phrase 'slippery slope'. Focus should instead be directed at the avoidance of adverse outcomes through means such as regulation, and at remedies should such outcomes materialise.

Of pertinence is the ethnographic finding of Sarah Franklin and Celia Roberts on the uptake of embryo research and new reproductive technologies in the UK. Contrary to the view of pre-implantation genetic diagnosis (PGD) as a socially destabilising technology in offering 'too much choice', their

---

[113]Stenvoll D. Slippery Slopes in Political Discourse, in *Political Language and Metaphor: Interpreting and Changing the World*, eds. T Carver and J Pikalo. London and New York: Routledge, 2008, pp. 28–40, p. 29. Stenvoll observes that metaphors are significant in politics because they have constitutive functions in that they contribute to the scene that set off a sequence that inevitably ends in tragedy (p. 37).

[114]Holm S and Takala T. High Hopes and Automatic Escalators: A Critique of Some New Arguments in Bioethics, *Journal of Medical Ethics* **33**, 1 (2007): 1–4.

study suggests a much more careful and thoughtful engagement with the difficulties presented by PGD.[115] In the light of this, the BAC considers effective regulation to be a means by which the determinism entailed in the 'slippery slope' argument could be countered. In the HA Consultation Paper, the BAC states:[116]

> *The UK is a country with one of the longest experiences with such a regime, first established under the Human Fertilisation and Embryology Act in 1990. This regime was in turn the result of a decade long process of deliberation and consultation since the publication of the Warnock Report. During the periods prior to and even after this regime has been established, there was concern that reproductive technologies may be misused for purposes such as eugenics. The 'slippery slope' argument was often raised as a basis for this concern. But for almost twenty years since its enactment, this legal and regulatory regime appears to have been effective in keeping reproductive technologies within acceptable ethical limits.*

Ironically, objectification and reification are — as we have seen — encompassed in countering the 'slippery slope' argument with regulation. As regulatory objects, chimeras and hybrids fall under the power of regulatory control. Earlier on, policy-makers discovered that the construction of the ethico-legal category of the 'pre-embryo'[117] is a means by which human embryo research could be kept off the slippery slope. Arguably, the ability to control 'pre-embryos' as 'things' underscored Dame Warnock's confidence in a legislative response to human embryo research.[118] As we have also seen, the distinction drawn between research and therapy, the segregation of motives, and the requirement of informed consent could all be regarded as ethical and legal 'technologies' by which the object of 'pre-embryo' is carved out of the social bedrock.

---

[115] *Ibid.*

[116] Bioethics Advisory Committee, Singapore, *Human-Animal Combinations for Biomedical Research: A Consultation Paper*, 8 January 2008, p. 29, para 60.

[117] That is, an 'embryo' from the point of creation up to time the primitive streak appears — which is approximately 14 days for a human embryo.

[118] Mulkay M. *The Embryo Research Debate: Science and the Politics of Reproduction.* Cambridge: Cambridge University Press, 1997, p. 149.

While the proposed models for chimeras and hybrids as presented in the HA Consultation Paper were successful in achieving some level of public consensus, as well as in moderating overtly adverse feedback, their acceptance was far from complete. From the results of the consultation, we see that some lay members of the public remain concerned that scientists are interested in creating live chimeric 'monsters' in spite of clear language to the contrary in the consultation paper. In the meeting with scientists, many of them appeared to have their own classificatory scheme.[119] Again, this is not a phenomenon that is confined to Singapore. Scientists in the UK initially used different definitions of chimeras and hybrids, making it necessary for the AMS and allied organisations to push for standardised meanings.[120] In Singapore, many scientists whom the BAC met with did not regard chimeras routinely created in the course of research to fall within the definition of human-animal combinations. Prior to the meeting, many of them considered human-animal combinations to relate only to cytoplasmic hybrid embryos.

SOME OBSERVATIONS ON NORM-MAKING AND SCIENTIFIC GOVERNANCE

Ethical and legal policies continue to revolve around scientific and technological change in the area of stem cell research and human-animal combinations. In fact, some human-animal combinations, such as transgenic mice, did not draw significant ethical or public attention until such combinations were categorically called into question following research proposals to create cybrids. Hence, these proposals became a trigger for re-opening ethical considerations in hESCR and cloning. As we have seen, developments in Singapore and the UK have largely been favourable to stem cell research. In contrast, cybrids represent the limits for stem cell research in some other countries, where the creation of such construct is prohibited.[121] These developments further illustrate recursivity in policy and norm-making on stem cell research. The inability to resolve debates over the moral status of a human embryo and related issues has contributed to repeated cycles of policy and normative evaluations, in a similar way to that observed in transnational

[119] Fieldnotes, 26 May 2009 (Ho WC).
[120] Fieldnotes, 25 September 2007 (Ho WC).
[121] The creation of a cytoplasmic hybrid embryo is prohibited under Section 5 of Canada's Assisted Human Reproduction Act, 2004.

norm-making on refugees and landmines.[122] In the UK, the 2008 amendments to the HFE Act of 1990 appear to have introduced a level of stability for the creation and use of certain types of human-animal combinations in research. However, a confluence of issue frames has begun so that there is now a less clear distinction between research involving human and non-human animals. The boundaries between ethical and regulatory provisions for research involving animals and those that apply to humans now require further consideration. Another possible consequence of this confluence is the proliferation of issue areas. For instance, transgenic animals previously considered to be ethically uncontentious have been earmarked for consideration by the AMS and other ethical bodies, such as the Danish Council of Ethics.[123]

The developments we have considered also suggest that norm-making cannot be said to be entirely global or local. At a broad level of analysis, the NAS may be said to have provided an early 'script' on human-animal combinations. The nature and rationale for their creation — and arguably their substantive meanings as such — cannot be understood independently of the ethical and regulatory proposals put forward by the NAS. This initial 'script' was adopted, but made general, by the ISSCR, as evident in its omission of the NAS's recommended prohibition of introducing human embryonic stem cells into non-human primate embryos. The guidelines of the ISSCR were a source of influence for the AMS in the UK, as well as the BAC in Singapore. However, the ISSCR and the BAC were subsequently influenced by the views of the AMS on cybrids. In these, we see 'international-local' and 'local-local' developments that are neither clearly 'global' nor 'local'.[124] Instead, the guidelines and documents of these various organisations appear as interlocking discursive spaces that enable, as well as facilitate, the recursivity in norm-making.

---

[122] Halliday TC. Recursivity of Global Normmaking: A Sociolegal Agenda, *Annual Review of Law and Social Sciences* **5** (2009): 263–289, p. 281.

[123] Danish Council of Ethics, Denmark, *Man or Mouse? Ethical Aspects of Chimaera Research,* 2007, p. 10.

[124] In certain areas such as the globalisation of US bankruptcy laws, it has been observed that the global is never entirely global since it always involves the globalisation of some local rule defined by relations of power in the global economy and polity. See Fourcade M and Savelsberg JJ. Introduction: Global Processes, National Institutions, Local Bricolage: Shaping Law in an Era of Globalization, *Law & Social Inquiry* **31**, 3 (2006): 513–519, p. 516. See also Santos BS. *Law and Globalization from Below: Towards a Cosmopolitan Legality.* Cambridge: Cambridge University Press, 2005.

Organisations such as the AMS, the BAC, and the ISSCR have important roles in filtering norms and regulatory processes. Quite unlike the emergence of a single consensual norm on insolvency that was articulated by the United Nations in 2004,[125] the filtering processes *vis-à-vis* stem cell research involving human-animal combinations have been multi-directional. Local norms are filtered through as global norms by the ISSCR. Global norms, as well as local norms, are filtered through by the AMS and the BAC. All three organisations have also been important in the standardisation of terminology and the prescription of ethical and regulatory forms as normative content. In addition, some level of epistemological continuity and connectivity could be observed in the way that chimeras and hybrids are associated with embryonic stem cell research. The HFE Act remains the overarching framework by which certain human-animal combinations such as cybrids are referenced in the UK. In Singapore, the ethical framework of 2002 is likely to assume a similar function. For both countries, continuity is maintained in linking research involving human-animal combinations to embryo research through a broadening of the normative 'script' for the latter. Technologically, iSCNT is regarded as a specialised application of SCNT , just as hSCNT is. This continuity in the enlargement of the 'script' implies that certain ethical obligations (such as the 14-day rule) that apply to embryo research should also apply in iSCNT .

Although there appears to be a general 'script' of shared norms on stem cell research involving human-animal combinations, there are also important differences, owing to differences in situated knowledge and institutional arrangements and practices. In Singapore, the government (through governmental organisations such as the BAC) remains a central force in spearheading normative change. Consequently, many of the normative requirements are also seen to have regulatory effect in practice.[126] In the UK, professional bodies such as the AMS have taken an active role in influencing the scope and character of legal and regulatory policies. This is perhaps most evident in a meeting that was arranged by the AMS and its allied organisations with members of the British Parliament. Nevertheless, a general 'script' is evident, arguably facilitated by state actors' recognition that shared norms are important to enable cross-border research collaborations. The central role of norms

[125]Carruthers BG and Halliday TC. Negotiating Globalization: Global Scripts and Intermediation in the Construction of Asian Insolvency Regimes, *Law & Social Inquiry* **31**, 3 (2006): 521–584, p. 539.

[126]Fieldnotes, 26 May 2009 (Ho WC).

in shaping regulatory policies and practices on stem cell research, both global and local, further indicates a broadening of analytical frames and stakeholders in scientific governance. As such, the shaping of science policies for such research can no longer be viewed as primarily a concern of technical experts, but is now inextricably linked to broader perceptions of order and progress.

10

How Will Future Bioethical Issues Engage Singapore?

John M. Elliott

Undertaking to comment on the future is to give hostages to fortune. Doing so with respect to a very small country is additionally hazardous, because a very small country has to be agile, nimble in changing policy, to survive. On the other hand, such countries must themselves gauge the future and anticipate, not merely react to events. At any rate, the challenge is likely to be interesting, and one can but try. Moreover, thinking about the future does allow a certain interesting latitude in the possibilities to be entertained. Inevitably, I consider future issues as they strike us in Singapore, but I aim to consider them generally rather than parochially.

The future seems clearly to contain the promise (or threat) of technical progress (or change). Science in general has a way of throwing up technologies that can challenge preconceptions, and this is certainly true in bioethics. Ethical principles, on the other hand, are not supposed to be changeable. 'Nimble' might be praise if applied to economic or research strategy, but it summons the wrong connotations (of opportunism, perhaps) if applied to ethics.

This might suggest that the best strategy for the present paper would be to extrapolate the present (presumptively fixed) ethics to the future, on the assumption that whatever the future might hold, we can at least be sure how it will be approached, ethically speaking. One could then select what seem to be future problematic areas of biology, and examine how the ethics might apply. I do this to some extent in what follows. However, I am strongly inclined to question the assumption that ethics are fixed. It seems to me that the most casual examination of history, and especially colonial history in Singapore, reveals a chasm in ethical principles between, say, 1950 and 2000. Indeed, the interest in modern biomedical ethics — as seen in various handbooks and histories for ethics review boards — to some extent reflects a reaction against a rather casual earlier attitude to what are now seen as important ethical issues.[1]

I therefore prefer to start with some discussion of the idea that ethics are changeable, and whether in fact there might be any way to anchor them. I then review the current ethical underpinnings of the BAC, before trying to anticipate the future and how we might react to it. I then ask whether the principles that have sustained the BAC for its first ten years will remain serviceable in the light of likely developments.

DO ETHICAL PRINCIPLES CHANGE?

In my view, it is a misconception to think that ethical principles remain unchanged as a fixed frame of reference for dealing with events as we encounter them. At most, I submit, we may reinterpret consistent principles to give them a new meaning, so that consistency of principle is maintained with a potentially very different set of actions. I have come to this view, in part, from having to consider the reactions of dissenting members of the public to issues raised in the course of deliberations by the BAC on issues before it.

For example, a common fear heard from critics of biomedical research, or some contentious aspects of it, is that a slippery slope is put in place. That as scientists gradually explore more and more possibilities, these will come to be acceptable. Often an accusation of scientific hubris is made, together with scepticism about the possible power of regulation to limit damage. However, I think the real but unarticulated fear is that *the standards of right and wrong will themselves change once we are used to the novelty.* A fresh point of departure will be established, allowing, after a decent lapse of time, a further descent.

Slippery slope criticisms were a clear theme in the feedback received when the BAC mooted the possibility of allowing and regulating embryonic

[1]See for example Amdur RJ. *Institutional Review Board: Member Handbook.* Sudbury, Massachusetts: Jones and Bartlett, 2003, pp. 9–21, and the chapter by Campbell, Chin and Voon in this book.

stem cell research. Several such slopes were featured, but one in particular was the fear that allowing research using Somatic Cell Nuclear Transfer (research cloning), in an attempt to create immuno-compatible cloned stem cells, would create a slippery slope leading to actual reproductive cloning, the creation of human clones. There was a clear view that the technology, once established, could and would be used willy–nilly for such undesired ends, and that we would not be able to regulate it once the genie was, as it were, out of the bottle.

It is noticeable here that the metaphor of descent is used, and I would like to challenge it. According to slippery slope criticisms we are to be thought of as descending, ethically speaking, in an undesired direction. Elevating metaphors, of escalators carrying us to new heights of ethical propriety, have not captured the imagination in any comparable fashion, if at all. Yet, one rebuttal of the slippery slope argument can start from the view that what we may first see as an undesirable change — *in vitro* fertilisation, for example — may, with familiarity, come to be seen as desirable. This is so because our moral standpoint changes *for the better*, or so we could maintain. What initially seems an affront to nature and an opening for exploitation becomes, in reality, a legitimate and viable means for the creation of children, whose provenance is utterly irrelevant to their claims on their families' affection and care.[2] It really is an elevator after all. Yet, and this is the point, such a prospect of apparent ethical relativism does in itself raise concerns. It is a very unsettling possibility for many people, because, as a general rule, we think of ethical standards as unchanging, and the basis on which we navigate the trials of life. A situation where a slippery slope, or an inexorable escalator, actually leads to changes in ethical standards undermines this security.

Such an ethical elevator, in my view, is to be seen in the post-colonial changes in ethical principles referred to earlier. Unreconstructed

[2]An analogous change over time, in Singapore, is the acceptance of the children of inter-racial marriages. Before independence it was not rare to encounter opposition along the lines of 'consider the children' when couples of different races intended marriage. (In the UK, such opposition was hilariously satirised and given a political twist by Michael Flanders and Donald Swann in their 1957 song 'Misalliance'). The high-minded assumption was that children should not be brought into a world where they would be exposed to inevitable slights and prejudices. Though real in a colonial context, times and attitudes have changed, thankfully. While racial prejudice as such is not yet extinct, it is now widely regarded as something socially and ethically unacceptable. The Flanders and Swann song is issued on CD (*At the Drop of a Hat* (1991) — EMI CDP 7974652). Performances and lyrics may be found with judicious on-line searching.

colonialists — Winston Churchill, for example — viewed the independence of India (in his terms, the 'loss' of India) as a step down a slippery slope which might lead, as indeed it did, to the independence of colonies generally, and the diminishment of the British Empire.[3] From a modern perspective, of course, the ethics of paternalism have thankfully been left behind — or rather, below — as we increasingly discard ideas of the relative superiority of civilisations or races.

I am not promoting here the idea that an escalator is somehow a positive argument for change, nor do I regard it as necessarily automatic. Like the slippery slope, I regard it as a questionable argument if produced as a reason for taking or not taking a course of action, for much the same reasons as given by Holm and Takala in their discussion of such arguments.[4] Rather, I am introducing it as an antidote to the idea that ethical change is bad because it is inevitably in the wrong direction. Change can be in either direction and good or bad depending on the facts of the case.

A further contribution to the unease felt in this matter, I believe, is that ethical standards, by their nature, are not objective, and it is difficult to find any basis for grounding them that can be agreed by everyone. They rather tend to be based on deeply held convictions and, as such, to be somehow unchallengeable, whether or not a cogent and objectively grounded argument can be produced. They are taken as premises, rather than contingent truths. Consequentialist or pragmatic arguments, by contrast, are apt to be dismissed as utilitarian and by implication unprincipled. The feelings of repugnance that some find in contemplating hybrid embryos or human clones, for example, may have no good grounding in an actual hazard or definable harm, but that does not stop them from being felt. Those who experience gut revulsion against some scientific claim or practice usually do not see the force of an argument that says their repugnance is an example of a naturalistic fallacy, needing support from a rational risk evaluation.[5]

---

[3] "The loss of India would be final and fatal to us. It could not fail to be part of a process that would reduce us to the scale of a minor power." Churchill, cited in Pearce RD. *The Turning Point in Africa: British Colonial Policy, 1938*–48. London: Cass, 1982, p 116.

[4] Holm S and Takala T. High Hopes and Automatic Escalators: A Critique of Some New Arguments in Bioethics, *Journal of Medical Ethics* **33**, 1 (2007): 1–4.

[5] For a classic appeal to repugnance *per se* as a guide to ethical judgment, see Kass L. The Wisdom of Repugnance: Why We Should Ban the Cloning of Humans. *New Republic* **216**, 22 (1997): 17–26.

Some people consider the element of moral relativity that I am identifying in this discussion to be more apparent than real. Such people might argue that essential ethical principles such as respect for individual autonomy have, in fact, been preserved, but manifest differently at different times. Others, however, do fear moral relativism lest it really be relative, and point to examples suggesting that the quality of public life is diminishing, for example, through increased reliance on surveillance and diminished reliance on personal morality. In my opinion, this fear is a fundamental reason behind the unease or opposition sometimes expressed to biomedical advances. It is arguably a kind of alienation based on powerlessness. It takes the form of hostility to perceived scientific power, and can produce anti-scientific attitudes, even among those whose everyday reliance on science and technology is manifestly considerable. Either way, it is clear that some consideration of future technical developments also has to seriously address the question of the ethical framework changes we might see, or need, because the latter are the means by which we deal with and evaluate the former.

Some philosophers, notably Rawls in his theory of justice,[6] have defended the possibility of providing some kind of an objective or universally compelling basis for ethics. However, ingenious as the 'Original Position' is, it does not, if I follow him, do the job Rawls thinks it does. It does not speak to the gamblers who reply they would sooner take their chances in an unequal world than endorse the view that not knowing one's destination in the world ensures a vote in favour of general rights.[7] Religions similarly, and unsurprisingly, tend to the view that there is a natural law or natural morality that, being of divine origin, is in a sense objective as well as unchallengeable, though this view is of course confounded with the issue of which religion is to be the determining one. This question is too large for extensive treatment here, though following Rachels[8] and de Waal[9] among other thinkers, and with some risk of committing a naturalistic fallacy, I prefer an approach that

[6] Rawls J. *A Theory of Justice*. Oxford: Clarendon Press, 1972.

[7] This is not an original criticism — it is of a kind general to libertarian ethicists such as Nozick (Nozick R. *Anarchy, State and Utopia*. New York: Basic Books, 1974).

[8] Rachels J. *Created from Animals: The Moral Implications of Darwinism*. Oxford: Oxford University Press, 1990.

[9] de Waal FBM. *Good Natured: The Origin of Right and Wrong in Humans and Other Animals*. Cambridge, Massachusetts: Harvard University Press, 1996. See also de Waal FBM. *Primates and Philosophers: How Morality Evolved*. Princeton, New Jersey: Princeton University Press, 2006.

starts with the role and evolutionary origin of morality and ethical thinking in human nature.

I think it is worth digressing slightly to put the essential point of this perspective briefly, as it is often misunderstood. There is now ample evidence from studies in evolutionary psychology and comparative ethology to indicate that desirable virtues, such as altruism, honesty, and cooperative behaviour, and their associated moral emotions, can be readily explained in principle as the result of natural selection operating at the level of individuals and their relatives. There may also be some effect at the level of societies, a view implicitly favoured by Darwin[10] and I believe first given explicit treatment by Sir Arthur Keith.[11] However, the modern consensus is that such group selection is unlikely to be a major effect. There has been a good deal of objection to the notion that such things as reciprocal altruism reflect 'real' or 'true' altruism, and not just some inferior 'biological altruism'. An early and clear account of this objection was given by Brown, who argued that this approach, like exchange theories in economics and social psychology, omitted a proper consideration of the psychological factors in genuine altruism.[12]

However, Brown's criticism, like many others, overlooks the crucial point that evolution operates on our feelings and emotions, such that these then tend to direct us into adaptive behaviours. Sincerity of feeling, genuineness of altruism, and feelings of guilt, obligation, or outrage at injustice, are just as plausibly selected characteristics of human nature as anything more mundane or physical, such as a liking for ripe fruit or for attractive members of the opposite sex. No young man, ardently contemplating the girl of his desires and affections, mutters to himself "good child-bearing hips" — though such considerations may indeed reflect why his preferences tend the way they do. What he experiences, and how he feels, are consequent on his paternal ancestors having felt similar desires, with resulting benefits in offspring. Those whose desires led in other directions — to older spouses, say — have left fewer descendants. So it is also with those whose sincere altruism and honest reliability led to their becoming valued members of society. Their sincerity is not devalued by the fact that it advantaged them in evolutionary terms, nor

[10] Darwin C. *The Descent of Man and Selection in Relation to Sex.* London: John Murray, 1871.
[11] Keith A. *A New theory of Human Evolution.* London: Watts, 1948.
[12] Brown R. *Social Psychology, the Second Edition.* London: Collier Macmillan, 1986.

does such advantage imply any ulterior motive. It is sincerity *qua* sincerity that was selected for.[13]

ETHICAL PRINCIPLES

Ethical principles, then, may not be as fixed as we would like to think, but may be grounded, though not determined, by natural preferences arising from evolved human nature. Given that, what principles do in fact activate the BAC?

The BAC accepts as a matter of course the obligation of beneficence enshrined in such standard reference documents as the Belmont Report.[14] In addition, it explicitly recognises autonomy, the importance of individuals being able to control their own lives as far as reasonably possible. It recognises proportionality — that restriction and regulation should be commensurate with possible harms. It recognises the importance of justice, a general principle of fairness and equity under the law, and, more widely, that costs and benefits in biomedicine should be equitably distributed and not be exploitative in character. It recognises what it calls sustainability, meaning the importance of not doing things that cannot be sustained or justified in the longer term. In addition, it recognises a principle of reciprocity. This is essentially the idea of the social contract, and is perhaps more strongly emphasised in collectivist rather than individualist cultures. In this contract the importance of autonomy, the emphasis on the rights of the individual, is counterbalanced by the rights of society, with a balance needed between the interests of the public and the needs of the individual. In fact, autonomy is constrained by all the other principles and is never absolute, in any society.[15]

The BAC is therefore far from unconventional, by international standards, in its general ethical outlook. It does, however, tend to have a somewhat

[13] For a brief but authoritative account of evolutionary psychology, see Dunbar R, Barrett L and Lycett J. *Evolutionary Psychology: A Beginner's Guide*. Oxford: Oneworld, 2007. For an account of the specific application of evolutionary principles to moral issues, see Wright R. *The Moral Animal: Evolutionary Psychology and Everyday Life*. New York: Pantheon Books, 1994.

[14] National Commission for the Protection of Human Subjects of Biomedical and Behavioral Research, USA. *Ethical Principles and Guidelines for the Protection of Human Subjects of Research*. Government Printing Office: Washington, DC, 1979.

[15] The most complete discussion of the ethical principles guiding the BAC is to be found in its Egg Donation Report: Bioethics Advisory Committee, *Donation of Human Eggs for Research*, 3 November 2008, p. 10.

consequentialist character. This is probably unavoidable given its role, which is to address and advise on ethical, legal, and social issues arising from biomedical research in Singapore; and given its position in a multi-religious and multicultural society. For reasons that I argue further below, a policy-level advisory body with such a remit and context is almost inevitably consequentialist.

To say this is not to imply that consequentialism in any of its various forms is undesirable, it is just an empirical statement about the nature of the BAC. In fact, Singapore generally could be described as, in this sense, consequentialist. Another word for it would be pragmatic, in the sense of being open to any practical solution to an issue. When faced with the business of deciding a course of action, a small country like Singapore tends to act with an eye to survival, and considers, very carefully, the consequences of its actions. This is understandable and indeed, if you live there, something to be deeply grateful for. It does not avoid any of the usual criticisms of consequentialist positions. The most fundamental — indeed, fatal — criticism, would seem to be that the question of why consequences are to be regarded as good or bad — desirable or not — is left unanswered. Consequentialism begs questions of right and wrong. Singaporeans are fond of such adages as 'not reinventing the wheel', which they use to mean that one should not replicate work already done elsewhere when finding solutions to problems. However, that one can recognise a serviceable wheel, a solution that works, when one sees it, is taken for granted; yet this is the nub of the matter. Those who suggest that Singapore should pick and choose best practice, from elsewhere or from its own diverse traditions, are simply ignoring the question of the criteria by which 'best practice' is to be recognised when found.[16] This is as true for ethics as for anything else.

Its small size is not the sole reason for a survival ethic to be so manifest in Singapore. It is a culturally and religiously diverse secular state.[17] There are

[16]GEM Anscombe, who originated the term 'consequentialism', and offered a scathing critique of it, anticipated this point in the context of borderline cases of good or bad acts; "... the consequentialist, in order to be imagining borderline cases at all, has of course to assume some sort of law or standard according to which this is a borderline case, Where then does he get the standard from?" Anscombe GEM. Modern Moral Philosophy, *Philosophy* **33**, 124 (1958): 1–19, p. 13.

[17]Ethnic communities in Singapore are not homogeneous, either linguistically or culturally. Indian communities include (for example) Tamils, Sikhs and Punjabis; Malay communities include the Bugis; the Chinese include numerous dialect groups together with the longer

numerous religions in Singapore, the figures for respective religious affiliations given in the 2000 census (the latest) being as follows:

Religious Affiliation	Buddhist	Christian	Muslim	**Taoist**	Hindu	Other	None
Pop.%	42.5	14.6	13.9	8.5	4.0	1.6	14.8

The Constitution of Singapore is secular, but it specifically confers on all citizens the right to practice and propagate their religion,[18] and provision is made for Syariah Law to apply to certain aspects of Muslim life, such as marriages and inheritance.[19] There is, in addition, specific provision to support religious tolerance by Act of Parliament, including provision for a Presidential Council for Religious Harmony.[20] In consequence, it follows that in deciding what might be generally acceptable to the Singapore public, and of national or public benefit, religious viewpoints are always considered and often explicitly noted. Conversely, however, policy is never a matter on which a particular religious view can prevail over a judgement based on wider secular public interest considerations. In practice, Singaporeans are accustomed to a pragmatic *modus vivendi* in which religious views are respected but are not the primary determinants of public policy.

established Peranakan or 'Straits Chinese' (a community of Chinese extraction regarded as indigenous to the Western Malay Peninsula and Singapore by long-established residence and with a distinctive culture); and there are numerous minority communities and many inter-ethnic marriages. Apart from the Malays, who are almost all Muslim, there is also wide diversity in religious affiliation cutting across these groups.

[18]Singapore Statutes: *Constitution of the Republic of Singapore.* Article 15 (1) states: "Every person has the right to profess and practise his religion and to propagate it."

[19]Singapore Statutes: *Administration of Muslim Law Act* (Cap. 3), Revised 2009.

[20]Singapore Statutes: *Maintenance of Religious Harmony Act* (Cap. 167A), Revised 2001. Among other provisions, it is not legal to publicly derogate other people's religions. Section 8 (1) states in part that "The Minister may make a restraining order against any priest, monk, pastor, imam, elder, office-bearer or any other person who is in a position of authority in any religious group or institution or any member thereof for the purposes specified in subsection (2) where the Minister is satisfied that that person has committed or is attempting to commit any of the following acts" — of which the first item is "(a) causing feelings of enmity, hatred, ill-will or hostility between different religious groups". It may be noted that the Minister cannot do this without giving prior notice of his intention, and with opportunity to be made for representations (and, usually, in practice, retraction).

In the case of bioethics, a similar philosophy prevails. This is partly because BAC members, being almost all Singaporeans, naturally tend to share the ethos of their country; but also because some version of what I am calling consequentialism is arguably the only reasonable position to take when in a position of recommending policy in a diverse nation. I have argued this position elsewhere,[21] but I address it again below, because it is relevant to answering the question posed in the title of this chapter.

Facts are important when deciding what is ethical, and it is an interesting question to what extent ethical disagreements exist or are compounded because people vary in their assumptions as to the facts of cases and the factual consequences of their position. However that may be, ethical principles, ideas of what is right and wrong and why, do not easily yield to argument, and disagreements can be profound and irreconcilable.

The difficulty of reconciliation is compounded when people take absolute positions of principle, rather than admitting compromises grounded in consequences. Even where religious issues are absent, it is quite possible to see a clear divide between arguments based on principle and those based on consequences; advocates of principle tend to regard consequences as irrelevant. For example, views on whether corporal punishment of children is acceptable or abusive can be sharply divided depending on whether they are determined deontologically, on some principle ('it is always wrong to hit a child', for example), or on a study of consequences ('is the effect of corporal punishment good or bad'). In vain may students of outcome produce evidence that corporal punishment has different effects in different cultures and circumstances.[22] Such evidence will be seen as irrelevant by those in the deontological camp. And I think it irrelevant to note that deontological or absolute principles are apt to have a religious origin, though they need not. What God has prohibited cannot be set aside by considerations of consequence — since that, according to Anscombe, is the very point of such prohibitions.[23] In practice, of course, few principles ever turn out to be really

[21] Elliott JM. Ethical Considerations in Human Stem Cell Research, in *Life Sciences: Law and Ethics*, eds. Kaan T and Liu ET. Singapore: Singapore Academy of Law and Bioethics Advisory Committee, 2006, pp. 54–75.

[22] For example, Lansford JE *et al.* Ethnic Differences in the Link between Physical Discipline and Later Adolescent Externalizing Behaviours, *Journal of Child Psychology and Psychiatry* **45**, 4 (2004): 801–812.

[23] "The prohibition of certain things simply in virtue of their description as such and such identifiable kinds of action, regardless of any further consequences, is certainly not the whole of the

absolute; there are widely (not universally) accepted exceptions to the rule that one should not kill. Nonetheless, many people will feel that their core ethical beliefs, taken as a coherent system and not picked apart individually, should indeed not be compromised by circumstances or by thought of consequences.

The fundamental argument, then, is this. In considering bioethical policy, I maintain that a government has to contend with incompatible deontological arguments or sources, and cannot, therefore, be other than consequentialist in its analysis of policy. It is a mistake to conflate what a policy mandates with what is right for each and every individual. There need be no presumption, on this view, that any individual is under any obligation to change what they think is right or wrong, merely because a government decision has been taken on some matter of ethical debate and public policy.

This may sound obvious, but it could be put this way: that there need be no presumption that governments have made the *right decision* in terms of the considerations relevant to the individual, even though when individuals petition the government, they appeal to just those considerations. Whether governments have made the right decision in terms of the considerations relevant to issues of *policy* is a different matter. Governments have to consider the range of ethical arguments and positions put forward by their citizenry, but they do not have to adjudicate these positions, and arguably should not do so. Deciding to allow embryonic stem cell research, or maintain the death penalty, or enforce an Act for the maintenance of religious harmony, for instance, are all decisions that will not satisfy all parties, and it may appear that a government has taken a principled stand that favours one or other side of the argument. But I submit that that is not what has occurred. The arguments that a government ought to find persuasive should be consequentialist in character, unless, indeed, the government is a theocracy, or otherwise driven by uniform ideological or religious conviction into deontological conformity. No doubt, policy still has to be based on the application of principles, to avoid the objection stated earlier, but it is in the application of principles to particular cases that consequences matter.

Hebrew Christian ethic; but it is a noteworthy feature of it... it is pretty well taken for obvious [among all moral philosophers since GE Moore] that a prohibition such as that on murder does not operate in face of some consequences. But of course the strictness of the prohibition has as its point *that you are not to be tempted by fear or hope of consequences*." Anscombe GEM. Modern Moral Philosophy, *Philosophy* **33**, 124 (1958): 1–19, p. 13 (emphasis in the original).

I do not wish to imply that this amounts to utilitarianism. It appears to me that a consequentialist focus is, rather, a kind of virtue ethics applied at the level of government. Government is supposed to be virtuous. It is supposed to have the best interest of the nation at heart, to be motivated by the intention to govern in the public interest (whatever that may be). Government is not supposed to be venal, and some of the strongest criticisms of governments arise if they are seen to allow nepotism and corruption in the public domain, even though these might arguably be the natural order of things at the level of individual lives. Everyone understands that family is special, that relatives enjoy special consideration at law and informally. No-one pretends that people should treat everyone the same in private as in public relationships; thus shopkeepers ought to sell at a fixed price to everyone, but generosity and a waiver of costs can be extended to relatives and friends with little risk of odium. Not so a government.

The public interest thus gets calibrated in terms of consequences, given some framework of principles and an assumption of virtuous government. If the assumption of virtue is missing, nothing a government says or does will be credible. But if the assumption is allowed, it can sustain ethical support for decisions argued on grounds of public goods even when these are not to the liking of particular citizens.

If one reviews the past decisions of the BAC, it is clear that they fit this model quite well. It takes as given the basic principles outlined above, such as a respect for individual autonomy and for some idea of a social contract, with a consequent need to strike a balance between the interests of individuals and society. The BAC probably strikes this balance somewhat more in favour of society over the individual than would be the case in Western democracies, since Asian societies tend to that emphasis. Be that as it may, it then proceeds to discover what people think, and to review what would be the probable consequences of particular possibilities in whatever domain it is interested in exploring. It then determines a decision in what it sees as the public interest.

Take the case of egg donation for research. It was found that while donation was undoubtedly a desirable good from the perspective of research, there were concerns with the potential for the exploitation of women, as might arise if payment for eggs went beyond a simple principle of reimbursement for expenses incurred. There was also concern arising from a potential inconsistency, if eggs were paid for, whereas organs for transplant were not. Making payment might have affected ongoing national debate about what is or should be the appropriate level of compensation for organ donors, always assuming that the principle of donation was not to be abandoned in favour of a market for organs. On top of this, there remained an ongoing disagreement within society as to the acceptability or otherwise of any research that specifically created embryos. In sum, the decision by the BAC in respect of reimbursement of costs to egg donors was considered in the light of consequences for other debates ongoing at that time. It was not decided simply by reference to deontological principles, such as a claim that it is in principle wrong for women to undergo invasive clinical procedures that are not for their personal medical benefit; or conversely, that it is a matter for individual women to decide about their own bodies, and not a concern of the state.[24]

To take another example, the BAC at one stage considered whether or not it should be possible for couples to choose the gender of their child, which is likely to become a reality thanks to sperm segregation technologies,[25] even if selective abortion or pre-implantation genetic diagnosis are ruled out. The BAC discussion was actually in the context of pre-implantation genetic diagnosis, but the eventual arguments that prevailed against permitting selection were that it might reinforce gender stereotyping, and could lead to an imbalance in gender ratio in the population,[26] as indeed appears to have happened in countries with some element of *de facto* gender selection, such as China.[27] This is a patently consequentialist analysis, applicable to any method of gender selection. Although I have suggested that the BAC favours this kind of approach, my real argument is that any policy-level body has to do something of the sort. So I assume that whatever is argued here may have some generality.

---

[24] Bioethics Advisory Committee, *Donation of Human Eggs for Research*, 3 November 2008.

[25] For example, MicroSort: "MicroSort® is an exclusive preconception sperm sorting technology, developed by the Genetics & IVF Institute (GIVF), and designed to increase the likelihood of conceiving a child of a particular gender. Genetics and IVF Institute is conducting a clinical trial using the MicroSort® technology. This research study uses MicroSort®'s patented technology, which is designed for the purpose of separating sperm either into those that primarily produce girls or those that primarily produce boys." Quoted from http://www.microsort.net/

[26] Bioethics Advisory Committee, *Genetic Testing and Genetic Research*, November 2005, para 4.46 and recommendation 10, pp. 35–36.

[27] See for example Gu BC. A New Era in China's Demographic Dynamics, in *The China Population and Labor Yearbook: The Approaching Lewis Turning Point and Its Policy Implications*, Vol 1, eds. Cai F and Du Y. Holland: Brill Academic Publishers, 2009, pp. 1–32.

EXPLORING THE ETHICAL FUTURE

How then will future bioethical issues engage us? One can ask what the future issues might be, and explore the ethical horizon over which problematic developments may become visible. One can then ask whether the principles and manner of engagement that have sustained the BAC for its first ten years will remain serviceable in the light of such developments.

It is a commonplace to note that advances in scientific technique confer power to transcend the natural limits that would otherwise prevail over what humans can do. One has only to consider the possibilities of surrogate motherhood, embryo enhancement, or the prevention of mitochondrial disease by the use of donor ooplasm to see at once that in the area of assisted reproduction alone, conventional notions of what can and should be done are easily challengeable by the pressure of developments in what is technically possible, or likely to become possible. If one extends the possibilities to include artificial life in various forms, the possibilities are really legion. Similarly, the advent of extensive and easily accessed electronic databases raises prospects of surveillance and state control, as well as epidemiological research, in ways that could not have been contemplated at a time when a lockable filing cabinet and a card index were cutting-edge technologies in database management.[28] The issues raised by the expansion of technique therefore tend to be seen as a challenge because they offer choices not previously available and therefore not considered, or sufficiently considered, in the ethical or legal frameworks that we inherit from our predecessors and from society generally.

ADVANCES IN SCIENTIFIC TECHNIQUES

The scientific techniques which will prove the most challenging I forecast to be those that throw into the greatest relief the conflicts between different

[28]Writing of his experience as Head of British Scientific Intelligence in World War II, RV Jones remarked that "Harold Blythe ... introduced me to the only type of filing system that I have ever found to work. This consists simply of box files into which papers of any size can be hurriedly deposited. I ended the war with more than four hundred files built up on his system, and I have continued to use it ever since." Jones RV. *Most Secret War: British Scientific Intelligence 1939–1945*. London: Hamish Hamilton, 1978, p. 144. It is clear that Jones — working in conditions of wartime secrecy — could never have used such a system had modern restrictions governing the acquisition and storage of sensitive data applied in his day. We may be duly thankful that they did not.

groups within society, or between the individual and society. Forecasting of this sort is rash, but it has to be done if we are to be prepared, and the unexpected may be upon us in a short time. Therefore let me briefly consider three areas of potential concern.

Enhancement

The first is enhancement. In my view, the distinction between the curative or restorative power of medicine, and its power to enhance and engineer desired outcomes that exceed nature, which is already hard to defend conceptually, is practically undermined by the activities of plastic surgeons and beauticians generally, and will become further undermined by advances in genetics that render the distinction between human and animal genes or DNA obsolete. We may need to reconsider our ethical landmarks in a possible world in which humans may not be clearly defined by a unique genome, but are, in effect, a fuzzy concept, with blurred genetic boundaries and ample opportunity and means for 'adjustment' at the edges.

Some of the conceptual difficulties in distinguishing enhancement from treatment become apparent if we consider a range of possibilities at or near the boundaries of the possible. Enhancement can be genetic or non-genetic. That is, it can be purely personal, or it can be extended to the germ-line and affect the offspring as well as the person treated. It can also occur via selection processes whereby a naturally occurring advantage is preferred through some process of embryo diagnosis and selection, or, more actively, through actual genetic engineering or other intervention intended to directly enhance an individual.

Consider the following set of theoretical possibilities in relation to ensuring offspring that are healthy and normal, and preferably smart, good looking, and of an equable temperament as well (but you can nominate your own wish list); and ask yourself how and where you would draw the line between enhancement and treatment. No assumptions have been made about legality, feasibility, or ethical propriety; this is just a consideration of possibilities for purposes of discussion. The possibilities have been ranked below in rough order of descending 'enhancement' but it is not a straightforward matter to do this.

1. Direct genetic modification to increase the likelihood of desired genetic characteristics in IVF embryos that you will then implant and bring to

term, with or without the use of a surrogate mother. These would be genetically modified 'designer babies' (so-called). This would certainly seem to count as a paradigm example of genetic enhancement.

2. Obtaining an oocyte and/or sperm from donors with desired traits and using IVF as the means to obtain and implant embryos with desired potential and thus bring to term babies that will by prior agreement be legally, though not genetically, your offspring, with or without the use of a surrogate mother and whether or not payment is made for the gametes used. Here this might not count as enhancement, since the embryo is not modified. On the other hand, this is not a treatment for infertility, but done voluntarily as a means to try and get a more desirably endowed child than one might have by natural means, which certainly counts as using a biomedical technique to improve on nature.
3. Selecting the gender of your child by selective abortion. As with scenario 2, the embryo, or foetus, is not changed; but here there is no presumption of a general social preference for a particular gender, so it is not clear that there is any sense in which enhancement can be said to occur, even though the procedure is not a treatment by any reasonable criterion. This also applies to scenario 4.
4. Selecting the probable gender of your child by the use of a pre-conception method, such as sperm selection (see above), if and when reliable techniques are available.
5. Avoiding genetic disease in your offspring by genetically modifying embryos using IVF. Here the avoidance of disease means this counts as treatment, but does the engineered perpetuation of this avoidance amount to enhancement in the sense of rendering treatment unnecessary? One might ask if environmental measures such as fluoridation of water supplies, count as enhancing the health and quality of life of children thus indiscriminately affected; if so the only difference in the genetic example would seem to be that it is genetic.
6. Treatment for mitochondrial disease by attempted permanent elimination of defective mitochondria, uniting a maternal somatic cell nucleus with an enucleated donated oocyte. All descendants in the matriline of a successfully treated female infant will obtain the benefit (and any unforeseen disadvantage). Male children in the matriline will have the benefit but will not transmit it. Similar considerations apply as for scenario 5, except that this might be considered less contentious because the genetic alteration only affects mitochondrial DNA and does not affect germ-line DNA.

7. Selection of a child that will be immuno-compatible with an existing sibling via selective implantation of a suitable embryo (so-called 'saviour siblings'). This technique exists and has been used in this way, for this purpose. It means the baby thus selected will be able to donate tissue to the sibling. Normally this is envisaged as a way of treating a pre-existing sick sibling, but it is also a possible precautionary measure to enable or facilitate future tissue donation in either direction. This is choosing an embryo, rather than enhancing one, but still, it is arranging matters to the advantage of the living by improving on what would occur if matters were left to natural reproductive processes. Ethically it is arguably choosing one child for the advantage of another, and not purely for its own sake. To do this undermines the autonomy of the donor sibling.
8. Avoiding genetic disease in your offspring by selective abortion or by not implanting affected embryos conceived and checked *in vitro*. This could be argued to be neither treatment nor enhancement, it is just avoiding the existence of a future child with a likely genetic disease. At the level of population genetics it is arguably enhancement.
9. Avoiding genetic disease or abnormality in your offspring by not having any. This represents an ethically uncompromised way of effecting population genetic improvement.
10. Accepting, or indeed welcoming, the procreative results of natural unprotected sex whatever they may be. The default baseline from which all other possibilities depart.

Only scenario 1 above would seem to be unambiguously enhancement of the individual, while scenario 2 would appear to be a form of positive eugenics. Moreover, what counts as an enhancement as distinct from a handicap or a disability is not a matter of general agreement, witness the arguments over whether deaf parents ought to be allowed — in principle at least — to use IVF as a means of positively selecting a congenitally deaf (i.e., 'enhanced') child.[29]

[29] A widespread complaint among disabled people is that efforts to eliminate disability in children inherently reflect a derogatory attitude towards those who already have such disabilities, as manifested, indeed, by the use of the word 'disabilities'. Some deaf individuals, moreover, deny they are disabled, or that their condition amounts to a handicap. Jonathan Glover has a good discussion of the need to separate the condition from the person, and so (he hopes) rebut the 'derogatory attitude' argument. See Glover J. *Choosing Children: Genes, Disability and Design*. Oxford: Oxford University Press, 2006, pp. 29–36 (Chapter 1, Disability and Genetic Choice).

Regardless of the evident difficulty of distinguishing enhancement from treatment, it is obvious that there are ethically contentious issues raised by virtually every one of these possibilities. One might think that scenario 10, being a state of nature, would be ethically uncontentious, but Harris has argued that we actually have an obligation to enhance people if we can.[30] Also relevant for the present discussion is the matter of whether research into any of possibilities (1–8) carries an ethical risk. It is one thing to propose or undertake an intervention. It is another to undertake research that might open up the possibility of the intervention. The BAC, in fact, is not an organisation set up to investigate or recommend on matters of clinical intervention — its remit is in matters relating to biomedical research.

A further difficulty in distinguishing treatment from enhancement surfaces if we consider the implication of a materialist and functionalist philosophy (I declare an interest here). Such a philosophy will see humans as defined by the functioning of their biological components, and not by some essential quality independent of such function. Embodiment of function may take various forms. Synthetic blood, artificial hearts, and incipient stem cell treatments are all taken to be successful by a criterion of satisfactory function. Therefore, so the argument runs, the provenance of the components is irrelevant provided they function equivalently to the natural case, and one does not become less human in any essential sense by incorporating extraneous artificial or even animal components into one's body, including one's nervous system. Function, not origin, defines us. Yet if this is accepted, the distinction between treatment and enhancement becomes even harder to maintain. We could end up in an absurd situation where one can ethically achieve a level of function superior to what one would have had naturally, an enhancement, by using an artificial component, provided one is first debilitated by some treatable condition.

The fundamental ethical issue is likely to be whether or not we can justify using our own judgement as a basis for determining what we will allow generally for others. Biomedical enhancement, if allowed, will create tensions not only between religious and non-religious perspectives, but also between the well-to-do and the impoverished, if enhancements and improvements are allowed but not equitably distributed. This is because some of the

[30] Harris J. *Enhancing Evolution: The Ethical Case for Making Better People.* Princeton, NJ: Princeton University Press, 2007.

greatest existing sources of social and economic inequality are biological inequalities, but these are presently seen as inevitable. To further create them artificially would be hard to justify in any ethical framework in which equality of opportunity is supported. As long as inequalities are seen as irreversible or unavoidable, one can focus on compensatory efforts. If they become directly changeable, or potentially so, they risk reinforcing social equality or inequality. We could describe the issue as the question of 'laboratory eugenics' — a deliberately provocative term, but not really an unfair one, I submit. Eugenics was defined originally by Francis Galton as "the study of all agencies under human control which can improve or impair the racial quality of future generations",[31] and later as "the conditions under which men [*sic*] of a high type are produced".[32] Laboratory eugenics I would simply define as the study of all agencies under human control which can improve the biological quality of future individuals by laboratory interventions that they themselves would desire (shades of Rawls). The qualification is an effort to ensure that autonomy is respected, but presumes to know what the autonomous individual would desire before they, being but an embryo, are able to desire it. Nonetheless, we are generally prepared to adopt such a position when it does not involve enhancement — as when we refuse to sanction the right of deaf parents to ensure deaf children — and it only seems contentious when we consider enhancement rather than treatment, and when it may be selective, not universal.

How will the BAC approach such an issue? To the extent that it has done so already, it has rejected enhancement under a sustainability principle, as exemplified in its views on gender selection discussed above. It is also inclined to be conventional in any matter where medical ethics are involved, and to be pragmatically cautious in allowing any innovation that would render the research environment out of alignment with that in other major research jurisdictions. On the other hand, informal observation suggests that many Singaporeans would not be averse to taking advantage of opportunities for cognitive or physical enhancement,[33] though some others would be strongly against it. Since the BAC is mainly concerned with issues

[31] Galton, F. *Inquiries into Human Faculty and its Development.* London: Macmillan, 1883, p. 24.

[32] Galton, F. *Inquiries into Human Faculty and its Development.* London: Macmillan (2nd ed.), 1893, p. 30.

[33] As for example, in various discussions with university students in evolutionary psychology classes, and on one occasion speaking to a group of pre-university students setting out, as a class

arising from research, it will likely not be in the front line of decisions on acceptability of medical enhancement; but it will have to consider its position should it become apparent that research with application in enhancement is contemplated. It is likely that the more basic the research, the more liberal the position the BAC will take, because it will view the advancement of basic research as a general good. To the extent that specific clinical applications become possible, it is likely that its position will be cautious or even hostile. In other words, it will tend to the rather classic view that fundamental science is generally to be supported, whereas issues of application — if they come to the BAC at all — will be viewed much more conservatively.

Primate Disease Models

A second area is the potential use of non-human primates as disease and treatment models. This raises issues of the ethics of research with primates. A recent paper by Sasaki *et al.* has demonstrated in principle that genetically modified marmosets can be reasonably freely produced and, most importantly, can pass their modification to their offspring.[34] The modification in this case was one resulting in marmosets that fluoresced under ultra-violet light, and the heritability of modifications was offered as a reason for thinking a potential primate model for human diseases was viable, with the presumed advantage of greater phylogenetic affinity. Though the authors do not say this, the advantage of phylogenetic affinity may be strongest when considering diseases of the central nervous system (CNS), but it is these very diseases in which the ethical dilemmas created by the use of primate models become most acute.

There is already a tendency to question the use of primates for research, and research with great apes is already banned in some countries, such as New Zealand and Sweden.[35] Marmosets, being phylogenetically more remote New World primates, may be a more borderline case, but the advantage of decreased ethical cost is somewhat offset by increased phylogenetic distance. This seems a clear example where, if the potential for clinical treatment

project, to explicitly consider the enhancement benefits of pre-implantation embryo selection or modification.

[34]Sasaki E *et al.* Generation of Transgenic Non-human Primates with Germline Transmission, *Nature* **459**, 7246 (2009): 523–528.

[35]For a disapproving comment, see the Editorial, Primate Rights? *Nature Neuroscience* **10**, 6 (2007): 669.

of deteriorating CNS diseases becomes a reality, the dilemma will become acute. Few would willingly forego (personally or for others) research into a treatment for Alzheimer's disease (say) in order to spare marmosets. Defenders of the rights of chimpanzees will become hard-pressed if the issue becomes practical rather than theoretical. It will quite clearly highlight the difference between a deontological position that says it is wrong to abrogate the rights we have conferred on great apes regardless of the advantages to humans, and those who will say it depends how big a gain humans would have from the research.

In addition, and more importantly, is the risk of humanisation of a primate CNS, either by the creation of transgenic (human/non-human) primates or by the use of disease models for treatments, or research into treatments, in which pluripotent human cells are introduced into developing primate brains. In the case of mice the addition of human stem cells to the brain has, rightly in my view, not been regarded as critical, but a far less relaxed attitude might be appropriate where primates are concerned. This possibility has already exercised scientists and ethicists,[36] and with the publication of the Sasaki *et al.* article, it seems closer to becoming a reality.

The issues raised here may become a concern of the BAC, which is currently considering human-animal combinations, including issues such as the possible humanisation of mental characteristics of animals by the addition of pluripotent stem cells. If some version of consequentialism prevails, there is likely to be a corresponding emphasis on regulation as the means of avoiding ethical or other harms. In the general case of research with animals, provision already exists for regulation to ensure animal welfare. But there is no strong tradition of animal rights in Singapore, though there is a considerable pet culture of a conventional kind, and a growing interest in wildlife conservation issues. Consequently, public opinion is not likely to be strongly against research with animals, including primates, if it is reasonably determined that it is fundamental to biomedical advances and does not harm the environment. There is some public misgiving about human-animal combinations, but the objections tend to be traditional ones focused on repugnance, human dignity, or on religious objections to creating modified embryos (cytoplasmic hybrids) which some appear to see as even worse than creating fully human embryos

[36] Greene M *et al.* Moral Issues of Human–Non-Human Primate Neural Grafting, *Science* **309**, 5733 (2005): 385–386.

for research. The response of the BAC is therefore likely to be to recommend research be allowed where the scientific case is strong, but with efforts to increase oversight commensurate to any perceived risk on a proportionality principle.

Artificial Intelligence and Synthetic Biology

The possible humanisation of primates in the pursuit of research, just discussed, is actually a subset of a much wider issue, that of the creation of synthetic life or life-like entities, in both biological and cognitive domains. Humans have been preoccupied for decades with the possibilities of constructed forms of machine intelligence[37] and, more recently, with constructed forms of life that are clearly biological. The provision of robotic carers, skilled workers, friends, counsellors, or even erotic companions[38] provides ample food for ethical debate even without the customary science fiction addition of fears about machines dominating or killing their human controllers.[39] An extended discussion of the ethics of robotics is well beyond the scope of this paper, but I want to make one specific point.

Many people cling determinedly to the view that a machine simply cannot be sentient or conscious, however well it simulates mental properties, a view given concrete expression by John Searle, but roundly attacked by proponents of artificial intelligence and philosophical functionalists generally.[40] However, let us assume for the moment that no robot can be sentient or conscious, however intelligent. This might provide a reason for confining the ethical debate on robots to their impact on humans, and not the impact of humans on them. If, however, engineered living organisms were allied to these abilities, one would immediately lose that rationale. One would cross a line into the construction of sentient and intelligent beings. Ethical issues would seem to arise both in considering the sentient entity and its rights and our obligations to it. The debate on such human-machine hybrids would have a strong resemblance to the debate on human-animal combinations.

[37] See http://www.cbc.ca/news/interactives/tl-robotics/ for a robotics timeline to 2007.

[38] The existence or actual marketing of robots with all these potentials is not science fiction. For a recent example, see http://www.truecompanion.com/

[39] For a general discussion of developments in robotics and predictions for the future, see Moravec H. Rise of the Robots — The Future of Artificial Intelligence, *Scientific American*, 18 (February 2008): 12–19.

[40] Searle JR. Minds, Brains, and Programs, *Behavioral and Brain Sciences* **3**, 3 (1980): 417–457. The article includes substantial critical peer review.

If we assume that robots might in fact one day be sentient, the issue of sentient robots becomes part of the wider issue of when it becomes simply unacceptable to create a sentient living or non-living thing with humanoid capabilities for purposes of research. This question needs to be considered whether the entity in question is entirely biological and living (albeit engineered); or whether it is some hybrid mix of biological and non-biological components; or even totally a machine. As long as these questions were being considered in the area of biomedical research, answers tended to be highly focused on eventual treatment possibilities. But the future in robotics and synthetic biology, separately or together, is not in the least confined to biomedical applications, and may be heavily driven by the marketplace.

An example of the issues that are likely to arise can be seen fairly compellingly by a consideration of the questions prompted by the appearance in the marketplace of sex robots (of either gender). These are not at present sentient, though they are responsive. They are not at present biological, though they are designed to closely simulate a living person. They can be regarded on one hand as the indulgence of the wealthy or sexually addicted, or on the other as a humane recourse for the socially incompetent or physically unattractive. A serious potential therapeutic function has been argued for this type of robot,[41] though the marketing techniques do not suggest that such considerations are foremost in the minds of salespersons. One might ask, would it make a difference if these robots were, in fact, conscious? And one could also ask, would it make a difference if they were, in fact, genetically engineered primates (looking just like humans, as do the robots) rather than electronically engineered robots with synthetic skin and organs?

Even if not conscious, it is difficult to imagine that genetically engineered living primate sex objects, however human-like, would engender anything other than the liveliest repugnance. If ever a case was to be made for the wisdom of repugnance, short of a revulsion at actually torturing or killing people, I imagine this might be it. One hardly knows where to begin enumerating the concerns. Since these creations would not be human, taboos against sex with animals would come into play. Since the relationship would be thoroughly exploitative, all the objections to exploitative relationships would arise, despite the fact that an engineered primate might not meaningfully be aware of exploitation, any more than is a trained and enthusiastic

---

[41] Levy, D. *Love and Sex with Robots: The Evolution of Human-Robot Relations*. New York: HarperCollins, 2007.

rat seeking out a landmine.[42] Any research technology capable of engineering a being of this kind in the first place would certainly be able to ensure there was no danger of inadvertent conception, but we might ask nonetheless why any person whose partner had made recent use of a sex robot might find their own sexual and affectionate impulses challenged. Such an action would dramatically undermine the exclusivity and personal character of intimate relationships.

Would these concerns be alleviated by the use of robots that were not biologically engineered, or were bio-mechanical hybrids? To some extent, perhaps. Any taboo on sexual intercourse with machines is probably not as strong as the one against animals, and no-one worries about exploiting machines, though machines that looked and acted like humans might be harder to ignore. But the problems created for actual relationships among humans would probably be almost the same. And of course, once the possibility of consciousness is admitted — whatever the actual makeup of the organism — then the issue of how one should treat such an entity becomes acute, and all the arguments against exploitation arise in great force.

The BAC is most unlikely to be relaxed or permissive about such matters. These issues will only come to its attention at the stage at which research ethics are considered, but it is at that point that some concern for the eventual use of research should be considered. In Singapore the principles of bioethical governance are not restricted to those funded publicly. Private enterprise is not left unregulated (or less regulated). Moreover, it is far from clear that the public interest would be served by the development of any form of biological entity that challenged the perceived autonomy and uniqueness of humans.

ARE OUR ETHICAL FRAMEWORKS SUSTAINABLE?

A salient consideration for the BAC, as for any advisory body, is the nature of the society in which it is situated, and how decision making is conducted. Formally a democracy born of democratic socialist roots, Singapore has attracted criticism as being, in reality, more controlled and less free than is compatible with democracies in more mature (or less controlled) societies.

[42] According to the BBC, rats have been trained to sniff out landmines and explosives. Dogs can also do this, but tend to set the mines off. BBC News. *Can Rats Help Clear Africa's Landmines?* 5 March 2010. http://news.bbc.co.uk/2/hi/8549681.stm

Its usual defence is that it has known what it is to be under occupation by two consecutive foreign powers (Britain and Japan, in that order), and that this experience, together with its unique blend of multiculturalism and the dictates of survival, has resulted in a strong independent survival ethic in which democratic freedoms are not regarded as paramount in comparison to the requirements of security and development. This accounts for why the BAC has been given a remit that includes the ethical, legal, and social issues arising from biomedical research and development and why it has been careful to both consider the views of the public, and educate them on the rationale for its recommendations. All the BAC reports are available on its website,[43] and all include as annexes the various letters and comments of the many public, professional, and religious groups consulted.

However, while it may be premature to assume that the ethical principles that have guided past BAC decisions are no longer adequate, they may need to be applied with some change of emphasis. The emphasis until now has been on the wellbeing of research participants rather than the implications of research for the wellbeing of society and the public at large. However, the challenges identified in this chapter are more the challenges arising from the products or consequences of research, and their effects on people and on society, rather than on participants. They are challenges to Justice and Sustainability. They may challenge the conventional civic assumptions of society. They do indeed reflect ethical, legal, and social issues arising from biomedical research and development. Perhaps, therefore, we will find that Justice and Sustainability become more clearly articulated and prominent principles. This would more clearly align thinking in research ethics with thinking in wider conservation issues, where the future is seen in terms of wider responsibilities to living systems generally and not just to those immediately affected or benefitting. It might also help make clear that basic research into fundamental principles should not automatically be seen as a threat to society just because some potential misuse can be envisaged. In a wired and interconnected world, the scope for imaginative application is considerable, but fear of unknown consequences has rarely in the past proved a good guide to the ethics of research.

[43] http://www.bioethics-singapore.org/

ANNEX A

Members of the Bioethics Advisory Committee and its Sub-Committees and Working Group

CURRENT MEMBERS OF THE BIOETHICS ADVISORY COMMITTEE

Chairman

Professor Lim Pin
University Professor, National University of Singapore
(since 2001)

Deputy Chairman

Professor Lee Hin Peng
Department of Epidemiology and Public Health
Yong Loo Lin School of Medicine, National University of Singapore
(since 2007)

Members

Professor Alastair Campbell
Chen Su Lan Centennial Professor in Medical Ethics
Director, Centre for Biomedical Ethics, Yong Loo Lin School of Medicine
National University of Singapore
(since 2007)

Mr Han Fook Kwang
Editor, *The Straits Times*
(since 2005)

Professor Kandiah Satkunanantham
Director of Medical Services, Ministry of Health
(since 2005)

Professor Eddie Kuo Chen-Yu
Professorial Fellow, Division of Communication Research
Wee Kim Wee School of Communication and Information
Nanyang Technological University
(since 2007)

Mr Charles Lim Aeng Cheng
Principal Senior State Counsel and Parliamentary Counsel
Legislation and Law Reform Division, Attorney-General's Chambers
(since 2005)

Mr Richard Magnus
Retired Senior District Judge
Subordinate Courts of Singapore, Representative to the ASEAN Intergovernmental Commission
on Human Rights
(2001 to 2006 [Deputy Chairman from 2005 to 2006]; 2009 to 2010)

Mr Nazirudin Mohd Nasir
Assistant Director, Office of the Mufti, Majlis Ugama Islam Singapura
(since 2007)

Professor Ng Soon Chye
Director, O & G Partners Clinic for Women and Fertility Centre
Gleneagles Hospital, Singapore
(since 2009)

Associate Professor Nuyen Anh Tuan
Department of Philosophy, National University of Singapore
(since 2005)

Associate Professor Patrick Tan Boon Ooi
Duke-NUS Graduate Medical School
Group Leader, Genome Institute of Singapore
(since 2007)

Professor Yap Hui Kim
Senior Consultant, Department of Paediatrics
Yong Loo Lin School of Medicine, National University of Singapore
(since 2007)

PAST MEMBERS OF THE BIOETHICS ADVISORY COMMITTEE

Associate Professor David Chan Kum Wah
Department of Philosophy, National University of Singapore
(January to May 2001)

Mr Jeffrey Chan Wah Teck
Head, Civil Division, Attorney-General's Chambers
(2001 to 2004)

Mr Cheong Yip Seng
Editor-in-Chief, Singapore Press Holdings
(2001 to 2004)

Associate Professor John Elliott
Department of Social Work and Psychology, National University of Singapore
(August 2001 to 2006)

Associate Professor Terry Kaan Sheung-Hung
Faculty of Law, National University of Singapore
(2001 to 2006)

Mr Ahmad Khalis bin Abdul Ghani
Member of Parliament, Hong Kah Group Representation Constituency
(2005 to 2006)

Professor Lee Eng Hin
Director, Division of Graduate Medical Studies
Yong Loo Lin School of Medicine, National University of Singapore
(2005 to 2008)

Professor Louis Lim
Executive Director, Biomedical Research Council
Agency for Science, Technology and Research
(2001 to 2002)

Ms Lim Soo Hoon
Permanent Secretary, Ministry of Community Development and Sports
(2001 to 2004)

Professor Edison Liu
Executive Director, Genome Institute of Singapore
(2003 to 2006)

Mr Niam Chiang Meng
Permanent Secretary, Ministry of Community Development
Youth and Sports
(2005 to 2006)

Professor Ong Yong Yau
Chairman, National Medical Ethics Committee
Ministry of Health
(2001 to 2004)

Professor Tan Chorh Chuan
Director of Medical Services, Ministry of Health (2001 to 2004)
Provost, National University of Singapore (2005 to 2006)
(2001 to 2006)

Mr Zainul Abidin Rasheed
Mayor, North East Community Development Council
(2001 to 2004)

MEMBERS OF THE PUBLICITY AND EDUCATION SUB-COMMITTEE

Current Members

Mr Han Fook Kwang (Chairman since 2005)

Professor Alastair Campbell (since 2007)

Associate Professor Denise Goh Li Meng, Senior Consultant, University Children's Medical Institute
National University Hospital (since 2007)

Professor Eddie Kuo Chen-Yu (since 2007)

Mr Nazirudin Mohd Nasir (since 2007)

Ms Bey Mui Leng, Head, REACH, Ministry of Community Development Youth and Sports (since August 2007)

Past Members

Mr Cheong Yip Seng (Chairman from 2001 to 2004)

Mr Ahmad Khalis bin Abdul Ghani (2005 to 2006)

Professor Louis Lim (2001 to 2002)

Professor Edison Liu (2003 to 2006)

Associate Professor Nuyen Anh Tuan (2005 to 2006)

Professor Ong Yong Yau (2001 to 2004)

Dr Clarence Tan, Chief Executive Officer, Health Sciences Authority (2001 to August 2003)

Mr Tan Yew Soon, Head, Feedback Unit, Ministry of Community Development and Sports (2001 to July 2004)

Mr Toh Yong Chuan, Head, Feedback Unit, Ministry of Community Development and Sports (July 2004 to August 2007)

Mr Zainul Abidin Rasheed (2001 to 2004)

MEMBERS OF THE SUB-COMMITTEE ON RESEARCH INVOLVING HUMAN PARTICIPANTS

Professor Lee Hin Peng (Chairman)

Professor Alastair Campbell

Mr Charles Lim Aeng Cheng

Mr Nazirudin Mohd Nasir

Associate Professor Patrick Tan Boon Ooi

Professor Yap Hui Kim

MEMBERS OF THE HUMAN EMBRYO AND CHIMERA RESEARCH WORKING GROUP

Mr Richard Magnus (Chairman in 2006 and from 2009)

Professor Lee Eng Hin (from 2006, Chairman from 2007 to 2008)

Professor Eddie Kuo Chen-Yu

Dr Lim Bing, Senior Group Leader, Genome Institute of Singapore

Mr Nazirudin Mohd Nasir

Professor Ng Soon Chye

Associate Professor Nuyen Anh Tuan

MEMBERS OF THE HUMAN GENETIC SUB-COMMITTEE (2001 to 2006)

Associate Professor Terry Kaan Sheung-Hung (Chairman)

Mr Jeffrey Chan Wah Teck (until December 2004)

Dr Samuel Chong, Department of Paediatrics, National University of Singapore

Dr Ong Toon Hui, Director, Elderly Development Division, Ministry of Community Development and Sports (until August 2003)

Professor Yap Hui Kim

Professor Chia Kee Seng, Department of Community, Occupational and Family Medicine, National University of Singapore (from September 2004)

Dr Denise Goh Li Meng (from September 2004)

Dr Lee Soo Chin, Consultant, Department of Haematology-Oncology National University Hospital (from September 2004)

Mr Charles Lim Aeng Cheng (from January 2005)

MEMBERS OF THE HUMAN STEM CELL RESEARCH SUB-COMMITTEE (2001 to 2005)

Mr Richard Magnus (Chairman)

Associate Professor David Chan Kum Wah (January to May 2001)

Associate Professor John Elliott (from August 2001)

Professor Lee Eng Hin

Ms Lim Soo Hoon

Professor Tan Chorh Chuan

Mr Zainul Abidin Rasheed

ANNEX B

Members of the International Panel of Experts

Professor Martin Bobrow (since March 2001)
Emeritus Professor of Medical Genetics
University of Cambridge, UK

Professor Bartha Maria Knoppers (since April 2003)
Professor of Law and Bioethics, Faculty of Medicine (Department of Genetics), and Director, Centre of Genomics and Policy
McGill University, Canada

Professor Bernard Lo (March 2001 to May 2002, and since February 2006)
Professor of Medicine and Director, Program in Medical Ethics
University of California, San Francisco, USA

Dr Thomas Murray (since May 2005)
President and CEO
The Hastings Center, USA

ANNEX C

Reports of the Bioethics Advisory Committee

No.	Title of Report	Date of Publication
1	Ethical, Legal and Social Issues in Human Stem Cell Research, Reproductive and Therapeutic Cloning	21 June 2002
2	Human Tissue Research	12 November 2002
3	Research Involving Human Subjects: Guidelines for IRBs	23 November 2004
4	Genetic Testing and Genetic Research	25 November 2005
5	Personal Information in Biomedical Research	7 May 2007
6	Donation of Human Eggs for Research	3 November 2008

ANNEX D

A Decade of Events that Shaped Bioethics in Singapore

Date	Events
2000	
Jan	US National Bioethics Advisory Commission published *Volume II* (*Commissioned Papers*) of its *Report and Recommendations on Ethical Issues in Human Stem Cell Research* (Volume I, which is the main Report, was published in September 1999)
Jan	World Health Organization (WHO) finalised its document, *Operational Guidelines for Ethics Committees that Review Biomedical Research*
Jan 1	Australian National Health and Medical Research Council (NHMRC) issued *Guidelines for Ethical Review of Research Proposals for Human Somatic Cell Gene Therapy and Related Therapies*
Jan 22	Iceland promulgated the Regulation on a Health Sector Database
Feb 23	UK House of Lords Select Committee on Science and Technology published a report on *Science and Society*
Mar	Australian NHMRC issued *Guidelines Under Section 95 of the Privacy Act 1988*

(*Continued*)

Mar 6	Japan — Sub-committee of Human Embryo Research, Bioethics Committee, Council for Science and Technology, published a report on *Human Embryo Research Focusing on the Human Embryonic Stem Cells*
Apr 6	UK Nuffield Council on Bioethics published a discussion paper entitled *Stem Cell Therapy: Ethical Issues*
Apr 9	Human Genome Organisation (HUGO) issued its *Statement on Benefit Sharing*, following principles given in its earlier *Statement on the Principled Conduct of Genetic Research* (1996)
Apr 13	Canada enacted the Personal Information Protection and Electronic Documents Act
May 11	Lithuania enacted the Law on Ethics of Biomedical Research
May 25	Iceland enacted the Biobanks Act
Jun	Singapore launched the Biomedical Sciences Initiative to develop the Biomedical Sciences cluster as one of the key pillars of the country's economy, alongside Electronics, Engineering and Chemicals (Phase 1 of the Initiative [2000–2005] focused on building a strong foundation in basic biomedical research)
Jun	Singapore Genomics Program launched (renamed Genome Institute of Singapore in June 2001)
Jun	UK Department of Health (Chief Medical Officer's Expert Group) published a report, *Stem Cell Research: Medical Progress with Responsibility*
Jun	US National Bioethics Advisory Commission published *Volume III* (*Religious Perspectives*) of its *Report and Recommendations on Ethical Issues in Human Stem Cell Research*
Jun 14	Japan — Bioethics Committee of the Council for Science and Technology published a report on *Fundamental Principles of Research on the Human Genome*

(*Continued*)

Jun 25	US President Bill Clinton announced the completion of an initial sequencing of the human genome by the Human Genome Project
Jul	ES Cell International Pte Ltd established in Singapore, with the aim to develop therapies from human embryonic stem cells.
Sep 6	Indian Council of Medical Research published *Ethical Guidelines for Biomedical Research on Human Subjects*
Oct	Singapore — Biomedical Research Council (BMRC) set up under the National Science and Technology Board (established in 1991 and later renamed the Agency for Science, Technology and Research [A*STAR] in 2002)
Oct	General Assembly of the World Medical Association amended the Declaration of Helsinki: Ethical Principles for Medical Research Involving Human Subjects
Oct 9	UK Medical Research Council issued a guidance document on *Personal Information in Medical Research*
Nov 14	European Group on Ethics in Science and New Technologies (EGE) published an opinion on the *Ethical Aspects of Human Stem Cell Research and Use*
Nov 17	Hong Kong enacted the Human Reproductive Technology Ordinance
Nov 30	Japan enacted the Act on Regulation of Human Cloning Techniques
Dec	**Singapore — Bioethics Advisory Committee (BAC) established to address the ethical, legal, and social issues arising from biomedical sciences research**
Dec 13	Estonia enacted the Human Genes Research Act
Dec 21	Australia enacted the Gene Technology Act
2001	
	Singapore — Ministry of Health (MOH) set up the National Disease Registries Office (renamed National Registry of Diseases Office in 2008)

(Continued)

	Japan — Ministry of Education, Culture, Sports, Science and Technology issued *Guidelines for Derivation and Utilization of Human Embryonic Stem Cells* and *Ethical Guidelines for Research on the Human Genome and Gene Analysis*
	Nepal Health Research Council published *National Ethical Guidelines for Health Research in Nepal*
Feb	Singapore — National Medical Ethics Committee published its *Ethical Guidelines for Gene Technology*
Feb	Australian Law Reform Commission and Australian Health Ethics Committee began joint inquiry into the protection of human genetic information
Feb 2	Finland enacted the Act on the Medical Use of Human Organs and Tissues
Feb 6	Iceland promulgated Regulations on the Keeping and Utilisation of Biological Samples in Biobanks
Feb 12	Human Genome Project announced the publication of a draft sequence and initial analysis of the human genome
Feb 28	UK Department of Health published its document on *Research Governance Framework for Health and Social Care*
Mar 20	UK House of Lords Select Committee published a report on *Human Genetic Databases: Challenges and Opportunities*
Apr	US National Bioethics Advisory Commission published a report on *Ethical and Policy Issues in International Research: Clinical Trials in Developing Countries*
Apr	UNESCO International Bioethics Committee (IBC) published a report on *The Use of Embryonic Stem Cells In Therapeutic Research*
Apr	HUGO Ethics Committee issued a *Statement on Gene Therapy and Research*

(*Continued*)

May 11	UK enacted the Health and Social Care Act 2001 (Sections 60 and 61 on patient information are of special relevance to the work of the BAC)
Jul	UK Department of Health issued Governance Arrangements for NHS Research Ethics Committees
Aug	US National Bioethics Advisory Commission published its Report and Recommendations on *Ethical and Policy Issues in Research Involving Human Participants*
Aug	Israel Academy of Sciences and Humanities Bioethics Advisory Committee published a report on the *Use of Embryonic Stem Cells for Therapeutic Purposes*
Aug 7	France and Germany requested the inclusion of a supplementary item in the agenda of the 56th session of the United Nations General Assembly (UNGA), namely, an **international convention against the reproductive cloning of human beings**
Aug 9	US President George Bush announced restrictions on human embryonic stem cell research using federal funds
Nov 9 to 30	Singapore — BAC conducted a public consultation on human stem cell research
Dec	German National Ethics Council published its opinion on *The Import of Human Embryonic Stem Cells*
Dec 4	UK enacted the Human Reproductive Cloning Act 2001
Dec 5	Japan — Ministry of Education, Culture, Sports, Science and Technology issued *Guidelines for Handling of a Specified Embryo*
Dec 12	United Nations General Assembly set up an **Ad Hoc Committee on an International Convention against the Reproductive Cloning of Human Beings**
2002	
	Council for International Organizations of Medical Sciences (CIOMS) published its revised *International Ethical Guidelines for Biomedical Research Involving Human Subjects*

(*Continued*)

	WHO published a *Handbook for Good Clinical Research Practice: Guidance for Implementation*
	Singapore — Biomedical Sciences Core Group of the Legal Service established
	General Assembly of the World Medical Association amended the Declaration of Helsinki: Ethical Principles for Medical Research Involving Human Subjects
Jan 2	Taiwan — Department of Health issued *Guidelines for Collection and Use of Human Specimens for Research*
Jan 24	Council of Europe — The *Additional Protocol to the Convention on Human Rights and Biomedicine, on Transplantation of Organs and Tissues of Human Origin* was introduced
Feb 13	UK House of Lords Select Committee published a report on *Stem Cell Research*
Feb 27 to Mar 13	Singapore — BAC conducted a public consultation on human tissue research
Mar	Malaysia — National Fatwa Council (the highest religious decision-making body) banned human cloning as it considered the procedure unnatural and against Islam
Apr 24	UK Nuffield Council on Bioethics published a report on *The Ethics of Research Related to Healthcare in Developing Countries*
May	UK Human Genetics Commission published a report, *Inside Information: Balancing Interests in the Use of Personal Genetic Data*
May	New Zealand Health Research Council issued *Guidelines on Ethics in Health Research*
May 23	Sweden promulgated the Biobanks in Medical Care Act
May 23	UK passed the Health Service (Control of Patient Information) Regulations
Jun 1	UK — The Health Service (Control of Patient Information) Regulations 2002 went into force (promulgated on May 23)

(*Continued*)

Jun 21	Singapore — BAC published a report, *Ethical, Legal and Social issues in Human Stem Cell Research, Reproductive and Therapeutic Cloning*
Jun 28	Germany enacted the Stem Cell Act
Jul	US President's Council on Bioethics published a report, *Human Cloning and Human Dignity: An Ethical Inquiry*, which recommended a four-year cloning ban
Jul	The Netherlands enacted the Embryo Act
Jul 17	Singapore — Deputy Prime Minister Dr Tony Tan announced the government's acceptance of BAC's recommendations on human cloning and stem cell research
Jul 18	UK House of Commons Science and Technology Committee published a report on *Developments in Human Genetics and Embryology*
July to Oct	UK Human Genetics Commission conducted a public consultation on the supply of genetic tests direct to the public
Aug	Singapore — Professor Ariff Bongso and his team at the National University Hospital announced that they had successfully grown a new human embryonic stem cell line on human feeder cells and cell nutrients, without using mouse cells
Sep 27	UK Nuffield Council on Bioethics published a report on *Genetics and Human Behaviour: the Ethical Context*
Oct 6	General Assembly of the World Medical Association adopted the **Declaration on Ethical Considerations Regarding Health Databases**
Nov 12	Singapore — BAC published a report on *Human Tissue Research*
Dec	HUGO Ethics Committee issued a *Statement on Genomic Databases*

(*Continued*)

Dec	Israel Academy of Sciences and Humanities Bioethics Advisory Committee published a report on *Population-Based Large-Scale Collections of DNA Samples and Databases of Genetic Information*
Dec 2	Launch of the Singapore Tissue Network (renamed Singapore Bio-Bank on 1 April 2010)
Dec 19	Australia enacted the Research Involving Human Embryos Act 2002 and Prohibition of Human Cloning for Reproduction Act 2002
2003	
	Completion of the sequencing of the human genome and 50th anniversary of the discovery of the DNA double helix
	WHO published a report on *Genetic Databases: Assessing the Benefits and the Impact on Human and Patient Rights*
	WHO published a report on the *Review of Ethical Issues in Medical Genetics,* which together with its 1998 *Proposed International Guidelines on Ethical Issues in Medical Genetics and Genetic Services* are intended for use by genetics professionals and policy makers to develop policies and practices appropriate for their own countries
	German National Ethics Council published an opinion on *Genetic Diagnosis Before and During Pregnancy*
Jan	UK Medical Research Council updated its document on *Personal Information in Medical Research* to include new guidance on the *Health and Social Care Act 2001: Section 60*
Mar	UK Human Genetics Commission published a report *Genes Direct: Ensuring the Effective Oversight of Genetic Tests Supplied Directly to the Public*
Mar 20	French National Consultative Ethics Committee for Health and Life Sciences published an opinion on *Ethical Issues Raised by Collections of Biological Material and Associated Information Data: "Biobanks", "Biolibraries"*

(*Continued*)

Apr	Singapore scientists determined the complete genetic code of the SARS virus
Apr 14	Human Genome Project completed more than two years ahead of schedule
Apr 24	French National Consultative Ethics Committee for Health and Life Sciences published an opinion, *Regarding the Obligation to Disclose Genetic Information of Concern to the Family in the Event of Necessity*
May	The final report of the Australian Law Reform Commission, *Essentially Yours: The Protection of Human Genetic Information in Australia*, tabled in the federal Parliament
May 11	Belgium enacted the Law on Research on Embryos *In Vitro*
May 23	Norway enacted the Act Relating to Biobanks
May 28	Denmark enacted the Act on a Biomedical Research Ethics Committee System and the Processing of Biomedical Research Projects
Jun	Inauguration of the International Society for Stem Cell Research (ISSCR)
Jun 10	Denmark amended the Law on Artificial Fertilization in Connection with Medical Treatment, Diagnosis, and Research (Research on Embryonic Stem Cells)
Jul	China — Ministry of Health issued *Guidelines on Human Assisted Reproductive Technologies*
Aug	Chinese researchers reported the generation of cytoplasmic hybrid embryonic stem cells through nuclear transfer of human somatic nuclei into rabbit oocytes
Sep	National University of Singapore (NUS) established its Institutional Review Board (IRB) to review, approve, and monitor the ethical aspects of all NUS research projects that involve human participants and human tissues, cells, and data
Sep 20	UK Nuffield Council on Bioethics published a report, *Pharmacogenetics: Ethical Issues*

(*Continued*)

Oct 16	**The International Declaration on Human Genetic Data** was adopted at UNESCO's 32nd General Conference (It was the 2nd international declaration in the field of bioethics to be adopted — after the **Universal Declaration on the Human Genome and Human Rights** in 1997)
Oct 27	Singapore — A*STAR's BMRC and the Juvenile Diabetes Research Foundation International (JDRF) signed an agreement to jointly establish a new S$5.2 m (US$3 m) funding programme to support stem cell research in Singapore (first research funding collaboration for BMRC with a philanthropic organisation and the first for JDRF in Singapore)
Oct 27 to 29	Singapore — International Stem Cell Conference organised by BMRC, A*STAR
Oct 29	Singapore — Biopolis (the centre of biomedical research in Singapore) officially opened
Nov	Singapore — National Advisory Committee for Laboratory Animal Research (NACLAR) established
Dec	Human-sheep chimeras with 7–15% of human cells in their livers produced to study the possibility of creating human tissues/organs in animals for transplantation
Dec 24	China — Ministry of Science and Technology and Ministry of Health promulgated the *Ethical Guidelines for Research on Human Embryonic Stem Cells*
2004	
	UNESCO published a document on *Human Cloning: Ethical Issues*
	Japan — Ministry of Education, Culture, Sports, Science and Technology revised its *Ethical Guidelines for Research on the Human Genome and Gene Analysis*
	Danish Council of Ethics published a debate outline on *The Beginning of Human Life and the Moral Status of the Embryo*

(*Continued*)

	General Assembly of the World Medical Association amended the Declaration of Helsinki: Ethical Principles for Medical Research Involving Human Subjects
Jan	US President's Council on Bioethics published a report on *Monitoring Stem Cell Research*
Jan	Malaysia's Ministry of Health published the second edition of the *Malaysian Guidelines for Good Clinical Practice* (the first edition was published in 1999)
Jan 6	Singapore — Human Organ Transplant Act revised
Jan 29	South Korea enacted the Bioethics and Biosafety Act
Feb 19	Italy promulgated the Regulation of Medically Assisted Reproduction
Mar	US President's Council on Bioethics published a report on *Reproduction and Responsibility: The Regulation of New Biotechnologies*
	German National Ethics Council published an opinion on *Biobanks for Research*
Mar	Israel amended its 1999 Law on Prohibition of Genetic Intervention (Human Cloning and Genetic Manipulation of Reproductive Cells)
Mar 29	Canada passed the Assisted Human Reproduction Act
Mar 31	European Union (EU) — Parliament and Council issued a *Directive (2004/23/EC) on Setting Standards of Quality and Safety for the Donation, Procurement, Testing, Processing, Preservation, Storage and Distribution of Human Tissues and Cells*
Apr	Singapore — National Healthcare Group's (NHG) Domain Specific Review Boards became operational
Apr 23	Finland updated the Medical Research Act
Jul	UNESCO's Division of the Ethics of Science and Technology published a document, *National Legislation Concerning Human Reproductive and Therapeutic Cloning*

(*Continued*)

Jul	Austrian Bioethics Commission at the Federal Chancellery published its report on *Preimplantation Genetic Diagnosis (PGD)*
Jul 5	Belgian Advisory Committee on Bioethics published an opinion on *The Free Availability of Genetic Tests*
Jul 12	Denmark issued a Ministerial Order on Information and Consent at Inclusion of Trial Subjects in Biomedical Research Projects
Sep	US — Ethics Committee of the American Society for Reproductive Medicine published its opinion on *Informed Consent and the Use of Gametes and Embryos for Research*
Sep	German National Ethics Council published its opinion on *Cloning for Reproductive Purposes and Cloning for the Purposes of Biomedical Research*
Sep 10	Singapore enacted the Human Cloning and Other Prohibited Practices Act
Sep 13	Irish Council for Bioethics published a guidance document on *Operational Procedures for Research Ethics Committees*
Oct	Japan — Ten Genetic-Medicine related Societies published the English edition of its August 2003 *Guidelines for Genetic Testing*
Oct 20	International Human Genome Sequencing Consortium published its scientific description of the completed human genome sequence, reducing the estimated number of protein-coding genes from 35,000 to only 20,000–25,000
Oct 28	Finland — Ministry of Social Affairs and Health issued a Government Decree on Gene Technology
Oct 29	Singapore — NACLAR issued *Guidelines on the Care and Use of Animals for Scientific Purposes*
Nov	HUGO Ethics Committee issued a *Statement on Stem Cells*
Nov 2	USA — California state passed the Proposition 71, an act which supported stem cell research and established the California Institute for Regenerative Medicine (CIRM)

(*Continued*)

Nov 6 to 7	Australian Health Ethics Committee hosted the 5th International Meeting of National Bioethics Advisory Bodies in Canberra (BAC participated for the first time)
Nov 9 to 12	Australia — 7th World Congress of Bioethics held in Sydney (BAC participated for the first time)
Nov 15	UK enacted the Human Tissue Act
Nov 18	UK Medical Research Council issued a guidance document on *Medical Research Involving Children*
Nov 21	New Zealand enacted the Human Assisted Reproductive Technology Act
Nov 23	BAC published a report on *Research Involving Human Subjects: Guidelines for IRBs*
Dec 24	South Korea enacted the Bioethics and Biosafety Act
2005	
	India — Ministry of Health and Family Welfare, together with the Indian Council of Medical Research and the National Academy of Medical Sciences (India), published *National Guidelines for Accreditation, Supervision and Regulation of ART Clinics in India*
	Finnish National Ethics Committees (National Advisory Board on Research Ethics [TENK], National Advisory Board on Health Care Ethics [ETENE], Sub-Committee on Medical Research Ethics [TUKIJA], Cooperation Group for Laboratory Animal Sciences [KYTÖ], National Advisory Board on Biotechnology [BTNK], and Board for Gene Technology [GTLK]) published a report on *Human Stem Cells, Cloning and Research*
	Nepal Health Research Council issued ***National Guidelines on Clinical Trials with the Use of Pharmaceutical Products***
	New Zealand Health Research Council updated its 2002 *Guidelines on Ethics in Health Research*

(*Continued*)

Jan	UK Medical Research Council issued a *Position Statement on Research Regulation and Ethics*
Jan 25	Council of Europe adopted the *Additional Protocol to the Convention on Human Rights and Biomedicine Concerning Biomedical Research* (the Convention on Human Rights and Biomedicine was signed in Oviedo in April 1997)
Feb 22	Malaysia — National Fatwa Council decided that research on embryonic stem cells derived from surplus embryos is allowed
Mar 1	Switzerland — Stem Cell Research Act became law
Mar 8	United Nations General Assembly adopted a non-binding **Declaration on Human Cloning**, calling for the prohibition of all forms of human cloning contrary to human dignity
Mar 14	UK Department of Health and the Association of British Insurers published a *Concordat and Moratorium on Genetics and Insurance*
Mar 17	UK Nuffield Council on Bioethics published a discussion paper on *The Ethics of Healthcare Related Research in Developing Countries* as a follow-up to its 2002 report on the same subject
Mar 24	UK House of Commons Science and Technology Committee published a report on *Human Reproductive Technologies and the Law*
Apr 5 to May 17	Singapore — BAC public consultation on genetic testing and genetic research
Apr 6	NUH-NUS Tissue Repository (formerly known as NUH Tissue Bank) officially launched
Apr 7	UK passed the Mental Capacity Act
Apr 24	UK Department of Health published its 2nd edition of *Research Governance Framework for Health and Social Care*

(*Continued*)

Apr 26	US National Academy of Sciences released *Guidelines for Embryonic Stem Cell Research*
May	US President's Council on Bioethics published a white paper on *Alternative Sources of Pluripotent Cells*
Jun	Australian federal government appointed a committee to review the **Prohibition of Human Cloning Act 2002 and the Research Involving Human Embryos Act 2002**
Jun 29	Irish Council for Bioethics published a report on *Human Biological Material: Recommendations for Collection, Use and Storage in Research*
Aug	Australia — Legislation Review Committee for the **Prohibition of Human Cloning Act 2002 and Research Involving Human Embryos Act 2002 released an *Issues Paper: Outline of Existing Legislation and Issues for Public Consultation***
Aug	German National Ethics Council published its opinion on *Predictive Health Information in Pre-employment Medical Examinations*
Sep	Opening of the Singapore Cord Blood Bank, Singapore's first and only public cord blood bank
Sep	Canadian Institutes of Health Research published the *CIHR Best Practices for Protecting Privacy in Health Research*
Sep 15	UK's Royal Society published a report, *Personalised Medicines: Hopes and Realities*
Oct	Singapore Stem Cell Consortium set up by A*STAR
Oct	Canadian Institutes of Health Research, Natural Sciences and Engineering Research Council of Canada, Social Sciences and Humanities Research Council of Canada updated its 1998 *Tri-Council Policy Statement: Ethical Conduct for Research Involving Humans.*
Oct 19	UNESCO's General Conference adopted the **Universal Declaration on Bioethics and Human Rights**

(*Continued*)

Nov	Portugal — National Council of Ethics for the Life Sciences published an opinion on *Stem Cell Research*
Nov 18	Singapore — BAC published a report on *Genetic Testing and Genetic Research*
Dec	First proof that human embryonic stem cells can develop into functional neurons when injected into the brains of foetal mice (Salk Institute for Biological Studies, US)
Dec	Swiss National Advisory Commission on Biomedical Ethics published its opinion on *Preimplantation Genetic Diagnosis*
Dec 13	EU initiated a new project *CHIMBRIDS* to analyse the scientific, ethical, philosophical, and legal issues raised by the use of chimeras and hybrids in research
Dec 19	Australia — Lockhart Review Committee submitted its report on the ***Legislation Review***: *Prohibition of Human Cloning Act 2002* and *Research Involving Human Embryos Act 2002* to the government
2006	
	Start of Phase 2 of Singapore's Biomedical Sciences Initiative (2006–2010), which focuses on building up the nation's translational and clinical research capabilities, while continuing to strengthen the foundation in basic sciences
	UNESCO published a document on *Ethics of Science and Technology: Explorations of the Frontiers of Science and Ethics*
	Japan — Ministry of Health, Labour and Welfare issued *Guidelines for the Clinical Research using Human Stem Cells*
Jan	Start of the Singapore Consortium of Cohort Studies, which is a long-term health study to discover how genetic, lifestyle, diet, and other environmental factors interact to impact our health

(*Continued*)

Jan	UK Human Genetics Commission published a report *Making Babies: Reproductive Decisions and Genetic Technologies*
Jan	Swiss National Advisory Commission on Biomedical Ethics published its opinion on *Research Involving Human Embryos and Fetuses*
Jan	Israel's Ministry of Health issued *Guidelines for Clinical Trials in Human Subjects*
Jan 1	Singapore — National Research Foundation (NRF) was set up to coordinate and fund research activities of different agencies, with an initial government funding of S$5 billion
Jan 17	UK — Academy of Medical Sciences published a report on *Personal Data for Public Good: Using Health Information in Medical Research*
Jan 31	US — CIRM Standards Working Group finalised its recommendations on the regulations for research which it funds
Feb 8	EU issued a *Commission Directive (2006/17/EC) Implementing Directive 2004/23/EC of the European Parliament and of the Council as Regards Certain Technical Requirements for the Donation, Procurement and Testing of Human Tissues and Cells*
Feb 24	Hinxton Group, an International Consortium on Stem Cells, Ethics and Law, issued its *Consensus Statement on Stem Cell Research*
Mar 15	Committee of Ministers of the Council of Europe adopted *Recommendations on Research on Biological Materials of Human Origin*
Mar 31	Singapore — Ministry of Health updated its *Directives for Private Healthcare Institutions Providing Assisted Reproduction Services* to include a section on research involving oocytes and embryos
Apr	Portugal — National Council of Ethics for the Life Sciences published an opinion on *Human Cloning*

(*Continued*)

Apr	New Zealand Ministry of Health updated its 2002 *Operational Standard for Ethics Committees*
Apr 1	Denmark amended the Act on a Biomedical Research Ethics Committee System and Processing of Biomedical Research Projects
Apr 6	Opening of the Laboratory of Stem Cell Biology under the Singapore Stem Cell Consortium
Apr 26	Czech Republic passed the **Act on Research on Human Embryonic Stem Cells and Related Activities and on Amendment to Some Related Acts**
May	UK Evaluation Forum, supported by the Academy of Medical Sciences, Medical Research Council, and Wellcome Trust, published a report on *Medical Research: Assessing the Benefits to Society*
May 23	Switzerland — Swiss Academy of Medical Sciences published ethical guidelines and recommendations on *Biobanks: Obtainment, Preservation and Utilisation of Human Biological Material*
May 27	Spain enacted the Law on Methods of Assisted Human Reproduction
Jun	Researchers at Japan's Kyoto University reported a world's first in converting mice somatic cells into cells with embryonic stem cell-like capabilities (induced pluripotent stem cells) by using four specific genes
Jun 14 to Jul 31	Singapore — BAC conducted a public consultation on the use of personal information in biomedical research
Jun 22	UK promulgated the Human Tissue Act 2004 (Persons who Lack Capacity to Consent and Transplants) Regulations
Jun 23	Philippine National Health Research System issued *National Ethical Guidelines for Health Research*
Jul	ES Cell International announced the creation of four human embryonic stem cell lines suitable for clinical use

(*Continued*)

Jul	Scientists from Newcastle University, Newcastle upon Tyne, UK, for the first time produced live mice using sperm derived *in vitro* from mouse embryonic stem cells
Jul 13	UK passed the Medicines for Human Use (Clinical Trials) Amendment Regulations
Aug 4 to 5	China hosted the 6th Global Summit of National Bioethics Advisory Bodies
Aug 6 to 9	China — 8th World Congress of Bioethics held in Beijing
Aug 11	UK — Nuffield Council on Bioethics published a report, *Genetic Screening: a Supplement to the 1993 Report by the Nuffield Council on Bioethics*
Sep	Singapore — Centre for Biomedical Ethics established in the Yong Loo Lin School of Medicine, National University of Singapore
Sep	British Medical Association issued a document, *Human Tissue Legislation: Guidance from the BMA's Medical Ethics Department*
Sep 8 to Dec 8	UK HFEA conducted a public consultation on donating eggs for research. A document entitled *Donating Eggs for Research: Safeguarding Donors* was issued.
Oct	Singapore — Biopolis Phase II (Neuros-Immunos) completed
Oct	Indian Council of Medical Research updated its 2000 *Ethical Guidelines for Biomedical Research on Human Participants*
Oct 19	Lithuania amended the 1996 Law on Donation and Transplantation of Human Tissues, Cells and Organs
Oct 24	EU issued a *Commission Directive 2006/86/EC Implementing Directive 2004/23/EC of the European Parliament and of the Council as Regards Traceability*

(*Continued*)

	Requirements, Notification of Serious Adverse Reactions and Events and Certain Technical Requirements for the Coding, Processing, Preservation, Storage and Distribution of Human Tissues and Cells
Oct 31	Malaysia — National Committee for Clinical Research issued *Guidelines on the Use of Biological Tissues for Research*
Nov	BAC and the Singapore Academy of Law published a monograph, *Life Sciences: Law and Ethics*
Nov	Irish Medicines Board published ***Guidelines for Pharmacogenetic Research***
Nov 17	French National Consultative Ethics Committee for Health and Life Sciences published an opinion on *Commercialisation of Human Stem Cells and Other Cell Lines*
Dec 5	Malaysia — Ministry of Health issued *Guidelines for Stem Cell Research*
Dec 12	Australia passed the Prohibition of Human Cloning for Reproduction and the Regulation of Human Embryo Research Amendment Act 2006
2007	
	The Netherlands amended the 2002 Embryo Act
Jan 11	China — Ministry of Health promulgated the *Regulation on Ethical Review of Biomedical Research Involving Human Subjects*
Jan 17	EGE published its *Opinion on the Ethical Aspects of Nanomedicine*
Jan 31	Singapore — BAC hosted the third meeting of the Ethics Working Party of the International Stem Cell Forum
Feb	German National Ethics Council published its opinion on *Predictive Health Information in the Conclusion of Insurance Contracts*

(*Continued*)

Feb 1	ISSCR issued *Guidelines for the Conduct of Human Embryonic Stem Cell Research* (dated 21 December 2006)
Feb 1 to 3	International Stem Cell Conference organised by the Singapore Stem Cell Consortium in conjunction with the International Stem Cell Forum Annual Meeting
Feb 6	US — Institute of Medicine and National Research Council of the National Academies released a workshop report on *Assessing the Medical Risks of Human Oocyte Donation for Stem Cell Research*
Feb 14	Estonia amended the Human Genes Research Act
Mar	Australian NHMRC updated its *National Statement on Ethical Conduct in Human Research*
Mar	Opening of the Singapore Stem Cell Bank, an initiative of A*STAR and hosted by the Singapore Stem Cell Consortium
Mar 16	Singapore — National Healthcare Group attained full accreditation from the Association for the Accreditation of Human Research Protection Programs, US
Mar 28	UK House of Commons published a report on *Government Proposals for the Regulation of Hybrid and Chimera Embryos*
Apr	Portugal — National Council of Ethics for the Life Sciences published an opinion on *Pre-implantation Genetic Diagnosis*
Apr 5	UK House of Commons Science and Technology Committee published a report on *Government Proposals for the Regulation of Hybrid and Chimera Embryos*
Apr 19	National Cancer Centre Singapore launched the Humphrey Oei Institute of Cancer Research
Apr 20	Official opening of the Singapore Institute for Clinical Sciences, with research programmes in Genetic Medicine, Hepatic Diseases and Metabolic Diseases

(*Continued*)

Apr 26	French National Consultative Ethics Committee for Health and Life Sciences published an opinion on *Biometrics, Identifying Data and Human Rights*
Apr 26 to Jul 20	UK HFEA public consultation on Hybrids and Chimeras: A Consultation on the Ethical and Social Implications of Creating Human/Animal Embryos in Research
May 7	Singapore — BAC published a report on *Personal Information in Biomedical Research*
May 9	Austrian Bioethics Commission at the Federal Chancellery published an opinion on *Biobanks for Medical Research*
May 20	Singapore — National Medical Ethics Committee published *Recommendations on Clinical Trials: Update Focusing on Phase 1 Trials*
Jun	Australian NHMRC published its revised *Ethical Guidelines on the Use of Assisted Reproductive Technology in Clinical Practice and Research*
Jun	Australian NHMRC published a report on *Challenging Ethical Issues in Contemporary Research on Human Beings*
Jun	UK — National Academy of Medical Sciences published a report on *Inter-species Embryos*
Jun	UK — Secretary of State for Health presented to the Parliament the *Government Response to the Report from the House of Commons Science and Technology Committee: Government Proposals for the Regulation of Hybrid and Chimera Embryos*
Jun	Portugal — National Council of Ethics for the Life Sciences published an opinion on the *Legal System for DNA Profile Databases*
Jun	Researchers from the Lifeline Cell Technology (a subsidiary of the International Stem Cell Corporation based in the US) and the Scientific Center for Obstetrics, Gynecology, and Perinatology of the Russian Academy of Medical Sciences

(*Continued*)

	reported the derivation of six human pluripotent stem cell lines from human parthenotes
Jun 13	Austrian Bioethics Commission at the Federal Chancellery published its Ruling on *Nanotechnology: A Catalogue of Ethical Problems and Recommendations*
Jun 20	EGE published its opinion on the ethical review of human embryonic stem cell research projects financed by the European Union: *Recommendations on the Ethical Review of hESC FP7 Research Projects*
Jun 29	Canadian Institutes for Health Research updated its *Guidelines for Human Pluripotent Stem Cell Research*
July	Taiwan — Department of Health issued *Human Research Ethics Policy Guidelines*
Jul 3	Spain enacted the Law on Biomedical Research
Aug	ISSCR published *Ethical Standards for Human-to-Animal Chimera Experiments in Stem Cell Research*
Aug	US — Ethics Committee of the American Society for Reproductive Medicine published a report on *Financial Compensation of Oocyte Donors*
Aug	The human embryonic stem cell line that South Korean researchers claimed was the result of somatic cell nuclear transfer was proven to be derived via parthenogenesis
Aug 1	ESHRE declared another voluntary five-year moratorium on human reproductive cloning
Aug 1	Hong Kong — 2000 Human Reproductive Technology Ordinance was updated
Sep	UK Medical Research Council announced funding of a research involving therapeutic cloning from a team at the North East England Stem Cell Institute, where women who choose to donate some of their surplus eggs for research

(*Continued*)

	would be partially reimbursed for part of their treatment costs
Sep	Stem Cell Network — Asia-Pacific (SNAP) established
Oct	UK HFEA published report on the findings of the consultation on *Hybrids and Chimeras*
Oct	Danish Council of Ethics and the Danish Ethical Council for Animals published a report, *Man or Mouse? Ethical Aspects of Chimaera Research*
Nov	Researchers from Japan's Kyoto University and University of Wisconsin-Madison in the US produced induced pluripotent stem cells produced from human adult skin cells
Nov	Researchers from the Oregon National Primate Research Center in the US succeeded in generating primate (rhesus macaque) embryos via a modified cloning approach and isolating two embryonic stem cells lines from these embryos, demonstrating proof-of-concept for therapeutic cloning in primates
Nov	Indian Council of Medical Research and the Department of Biotechnology published *Guidelines on Stem Cell Research and Therapy*
Nov	Researchers from Japan's Kyoto University created induced pluripotent stem cells from human skin cells without the use of a gene that could increase risk of cancer
Nov	Two teams of Chinese researchers reported the creation of parthenogenetic human embryonic stem cell lines
Nov 7 to Jan 7 2008	Singapore — BAC conducted a public consultation on the donation of human eggs for research
Nov 12	Singapore enacted the National Registry of Disease Act
Nov 13	UK MRC issued an ethics guide on *Medical Research Involving Adults Who Cannot Consent*
Nov 28	ES Cell International announced the production of the world's first clinical-grade stem cell lines

(*Continued*)

Dec	Singapore — Ministry of Health published its *Operational Guidelines for IRBs*
Dec	UK Human Genetics Commission published *More Genes Direct: A Report on Developments in the Availability, Marketing and Regulation of Genetic Tests Supplied Directly to the Public* (an update of its 2003 report ***Genes Direct: Ensuring the Effective Oversight of Genetic Tests Supplied Directly to the Public***)
Dec	Researchers from the International Stem Cell Corporation and the Scientific Center for Obstetrics, Gynecology, and Perinatology of the Russian Academy of Medical Sciences reported the derivation of four HLA homozygous stem cell lines from human parthenotes (one cell line has the potential to be a source of therapeutic cells that will minimise immune rejection after transplantation into hundreds of millions of individuals of differing sexes, ages, and racial groups)
2008	
Jan	Stem Cell Society, Singapore, registered
Jan	UK HFEA granted two one-year licences to create human-animal hybrid embryos for research — one to King's College London to study neurodegenerative diseases such as Alzheimer's disease, Parkinson's disease, and Spinal Muscular Atrophy and the other to Newcastle University to study the processes in nuclear reprogramming.
Jan 8 to Mar 10	Singapore — BAC conducted a public consultation on human-animal combinations for biomedical research
Jan 15	Inauguration of the Singapore Immunology Network, an initiative of BMRC, A*STAR
Jan 21	Singapore amended the Human Organ Transplant Act
Feb	A team of researchers from Japan's Kyoto University reported for the first time the generation of photoreceptor-like cells from mouse, monkey, and human embryonic stem cells under defined culture conditions

(*Continued*)

Mar 27	UK's first human-animal (cow) hybrid embryos created by researchers from Newcastle University
Apr 11	Hinxton Group issued its *Consensus Statement: Science, Ethics and Policy Challenges of Pluripotent Stem Cell-Derived Gametes*
Apr 11	German Parliament passed amendment to Stem Cell Act
Apr 23	Irish Council for Bioethics published an opinion on *Ethical, Scientific and Legal Issues Concerning Stem Cell Research*
May	First transgenic monkey (rhesus macaque) model of Huntington's disease produced (Emory University, US)
May 19	Austrian Bioethics Commission at the Federal Chancellery published a report on *Umbilical Cord Blood Banking*
May 19	UK General Medical Council issued a guidance document on *Consent: Patients and Doctors Making Decisions Together*
May 21	US enacted the Genetic Information Nondiscrimination Act
May 29	French National Consultative Ethics Committee for Health and Life Sciences published an opinion on *The "Personal Medical Record" and Computerisation of Health-Related Data*
Jun	Health Research Council of New Zealand published *Guidelines for Ethics Committee Accreditation*
Jun 5	South Korea revised its 2004 Bioethics and Biosafety Act
Jun 9	UK — Biotechnology and Biological Sciences Research Council published a report on *Synthetic Biology: Social and Ethical Challenges*
Jun 13	Association of British Insurers announced the extension of the moratorium on the use of genetic test results by UK insurance companies until 2014
Jun 30	UK HFEA granted a 12-month licence to the Clinical Sciences Research Institute, University of Warwick, to

(*Continued*)

	create human-pig hybrid embryos to study cardiomyopathy — the third human-animal hybrid embryo licence to be granted
Aug	A team of researchers at the Harvard University have produced 20 disease-specific stem cell lines using the induced pluripotent stem cell technique
Sep 1 to 2	French National Consultative Ethics Committee for Health and Life hosted the 7th Global Summit of National Bioethics Advisory Bodies
Sep 4 to 9	Croatia — 9th World Congress of Bioethics held in Rijeka
Sep 5	US National Academy of Sciences published *Amendments to the Guidelines for Human Embryonic Stem Cell Research*
Sep 15	Singapore passed the Mental Capacity Act
Oct 17	Singapore — Opening of Fusionopolis, the epicentre for infocomm technology, media, and sciences
Oct 22	General Assembly of the World Medical Association amended the Declaration of Helsinki: Ethical Principles for Medical Research Involving Human Subjects
Nov 3	Singapore — BAC published a report on *Donation of Human Eggs for Research*
Nov 13	UK — Human Fertilisation and Embryology Act 2008 enacted to amend the Human Fertilisation and Embryology Act 1990 and the Surrogacy Arrangements Act 1985, and to make provision on parentage
Nov 27	Council of Europe — The *Additional Protocol to the Convention on Human Rights and Biomedicine, concerning Genetic Testing for Health Purposes* was signed
Dec 3	ISSCR released its *Guidelines for the Clinical Translation of Stem Cells*
Dec 8	UNESCO IBC published a report on *Consent*

(*Continued*)

2009	
	CIOMS published its revised *International Ethical Guidelines for Epidemiological Studies*
Jan	Researchers from the Genome Institute of Singapore reported a new method to create iPS cells, by using the orphan nuclear receptor Esrrb in conjunction with two other transcription factors (Oct4 and Sox2)
Jan 23	US FDA approved the world's first clinical trial of human embryonic stem cell-based therapy using GRNOPC1 for patients with acute spinal cord injury (Geron Corporation)
Feb	The ability of animal eggs to fully reprogramme human somatic cells is being questioned by a team of researchers studying the reprogramming process of human somatic nuclei using both human and animal eggs
Feb	Researchers from the Max Planck Institute for Molecular Biomedicine in Münster, Germany, showed that cells can be reprogrammed to pluripotency by using just one of the standard four genes
Mar	Swiss National Advisory Commission on Biomedical Ethics published its opinion on *Research Involving Children*
Mar	Final outcome of EU CHIMBRIDS project — publication of the book, *CHIMBRIDS: Chimeras and Hybrids in Comparative European and International Research*
Mar	American Society for Reproductive Medicine published a report on *Donating Spare Embryos for Stem Cell Research*
Mar	Two teams of researchers (led by Andreas Nagy, of the Samuel Lunenfeld Research Institute at Mount Sinai Hospital in Toronto, Canada, and Keisuke Kaji, of the University of Edinburgh, UK) reported a new approach to creating iPS cells that does not involve viruses and would also allow for the genes that are inserted to trigger cell reprogramming to be removed
Mar	Researchers from the Morgridge Institute for Research, University of Wisconsin-Madison, reported the creation of iPS cells that are free of viruses and exotic genes

(*Continued*)

Mar 16	Austrian Bioethics Commission at the Federal Chancellery published its report on *Research on Human Embryonic Stem Cells*
Apr	Transgenic dogs (beagles) which could be used to study human genetic diseases produced (Seoul National University, South Korea)
Apr	Researchers from the Scripps Research Institute in La Jolla, California, have for the first time reprogrammed adult skin cells into an embryonic-like state using proteins instead of genes
May	UK — Royal Academy of Engineering published a report on *Synthetic Biology: Scope, Applications and Implications*
Jun	Chinese researchers have successfully reprogrammed pig skin and bone marrow cells to create the first pig pluripotent stem cells
Jun	A team of researchers from China's Sun Yat-Sen University and the University of South Florida and University of Connecticut in the US reported the birth of mice developed from parthenogenetic embryonic stem cells
Jun 2	ISSCR published its Ethics Report on *Interspecies Somatic Cell Nuclear Transfer Research*
Jun 9	UNESCO IBC published a report on *Human Cloning and International Governance*
Jun 23	UK Academy of Medical Sciences, Wellcome Trust, Medical Research Council, and the Science Media Centre published a booklet *Hype, Hope and Hybrids: Science, Policy and Media Perspectives of the Human Fertilisation and Embryology Bill,* presenting ways in which the scientific community helped inform the debate on hybrid embryos
Jul	Two separate teams of Chinese researchers report the creation of live mice from iPS cells
Jul 7	UK House of Lords Science and Technology Committee published a report on *Genomic Medicine*

(*Continued*)

Jul 7	US National Institutes of Health's new *Guidelines on Human Stem Cell Research* became effective
Jul 13	Austrian Bioethics Commission at the Federal Chancellery published its opinion on *Assistive Technologies Ethical Aspects of the Development and Use of Assistive Technologies*
Jul 27	Health Sciences Authority of Singapore announced the expansion and upgrading of its Cell Processing Laboratory into a state-of-the-art Cell Therapy Facility to advance the field of cellular therapy
Aug	Japan — Ministry of Education, Culture, Sports, Science and Technology revised its *Guidelines for Derivation and Utilization of Human Embryonic Stem Cells*
Aug 27	US FDA placed on clinical hold Geron Corp's GRNOPC1, pending review of new nonclinical animal study data (cysts were found at treatment sites in animals)
Aug 20	Malaysia's Ministry of Health launched guidelines on stem cell therapy and research: *Guidelines for Stem Cell Research and Therapy, National Standards for Stem Cell Transplantation: Collection, Processing, Storage and Infusion of Haemopoietic Stem Cells and Therapeutic cells (July 2009)* and *National Standards for Cord Blood Banking and Transplantation (January 2008)*
Sep 15	UK Human Tissue Authority's revised *Codes of Practice* came into force
Oct	International Federation of Gynecology and Obstetrics (FIGO) Committee for the Ethical Aspects of Human Reproduction and Women's Health released a report on *Ethical Guidelines Concerning Cytoplasmic Animal-Human Hybrid Embryos*
Oct 15	French National Consultative Ethics Committee for Health and Life Sciences published an *Opinion on Ethical Issues*

(*Continued*)

	in Connection with Antenatal Diagnosis: Prenatal Diagnosis (PND) and Preimplantation Genetic Diagnosis (PGD)
Oct 27	Australian NHMRC issued *Guidelines for Health Practitioners in the Private Sector on the Use and Disclosure of Genetic Information to a Patient's Genetic Relatives under Section 95AA of the Privacy Act 1988*
Oct 30	US FDA and Geron Corp reached an agreement concerning the Phase I clinical trial of GRNOPC1 (further pre-clinical studies to be done — Geron expects the clinical trial to re-start in the third quarter of 2010)
Nov	UK Human Genetics Commission published a report, *Nothing to Hide, Nothing to Fear? Balancing Individual Rights and the Public Interest in the Governance and Use of the National DNA Database*
Nov	Irish Council for Bioethics published an opinion on *Biometrics: Enhancing Security or Invading Privacy?*
Nov 10	UK Academy of Medical Sciences launched a new study to examine the use of animals containing human material in scientific research
Nov 17	EGE published an opinion on the *Ethics of Synthetic Biology*
Nov 24	US President Barack Obama established the Presidential Commission for the Study of Bioethical Issues
2010	
Jan	Researchers from Stanford University School of Medicine reported success in transforming mouse skin cells into functional nerve cells without going through the pluripotent stem cell stage
Feb	Researchers from the Genome Institute of Singapore discovered that a genetic molecule (Tbx3) could improve the quality of iPS cells created

(*Continued*)

Mar	Malaysian Medical Council issued *Guidelines on Stem Cell Research and Stem Cell Therapy*, which was adopted by the Council on 13 October 2009
Mar 24	UK passed the Human Fertilisation and Embryology (Disclosure of Information for Research Purposes) Regulations
Mar 27	UK Academy of Medical Sciences commissioned a leading market research company to explore public views on research using animals containing human material
Apr	Researchers from UK's Newcastle University announced success in a new IVF technique for preventing mitochondrial disease (embryo nuclear transfer)
Apr 20 to Jul 13	UK Nuffield Council on Bioethics public consultation on *Human Bodies in Medicine and Research*
May	Researchers at the J Craig Venter Institute announced that they have created a synthetic cell
May 20	US President Obama requested the Presidential Commission for the Study of Bioethical Issues to study the implications of advances in the field of **synthetic biology**
May 26	US National Research Council and Institute of Medicine of the National Academies released its updated *Guidelines for Human Embryonic Stem Cell Research*
Jul 26 to 27	Singapore — 8th Global Summit of National Bioethics Advisory Bodies (hosted by BAC and MOH, and supported by the WHO and European Union)
Jul 28 to 31	Singapore — 10th World Congress of Bioethics (organised by BAC, MOH, NUS Centre for Biomedical Ethics, and the Singapore Medical Association)
Jul 28	Launch of a permanent Bioethics Exhibition at Science Centre Singapore (collaborative project of Science Centre Singapore, BAC, and NUS Centre for Biomedical Ethics)

References

A

Aalto-Setälä K, Conklin BR and Lo B. Obtaining Consent for Future Research with Induced Pluripotent Cells: Opportunities and Challenges, *Public Library of Science (PLoS) Biology* **7**, 2 (2009): e1000042.

Amdur RJ. *Institutional Review Board: Member Handbook*. Sudbury, Massachusetts: Jones and Bartlett, 2003.

American College of Obstetricians and Gynecologists, Committee on Ethics. Committee Opinion Number 321: Maternal Decision Making, Ethics, and the Law, *Obstetrics and Gynecology* **106** (2005): 1127–1137.

Annas GJ, Andrews LB and Isasi RM. Protecting the Endangered Human: Toward an International Treaty Prohibiting Cloning and Inheritable Alterations, *American Journal of Law & Medicine* **28**, 2&3 (2002): 151–178.

Anscombe GEM. Modern Moral Philosophy, *Philosophy* **33**, 124 (1958): 1 19.

Appelbaum PS, Roth LH and Lidz CW. The Therapeutic Misconception: Informed Consent in Psychiatric Research, *International Journal of Law and Psychiatry* **5**, 3–4 (1982): 319–329.

Appelbaum PS, Roth LH, Lidz CW, Benson P and Winslade W. False Hopes and Best Data: Consent to Research and the Therapeutic Misconception, *Hastings Center Report* **17**, 2 (1987): 20–24.

Appelbaum PS, Lidz CW and Grisso T. Therapeutic Misconception in Clinical Research: Frequency and Risk Factors, *IRB* **26**, 2 (2004): 1–8.

B

BBC News. Can Rats Help Clear Africa's Landmines? 5 March 2010.

Balaji S. A Legislative Framework for Stem Cell Research in Singapore, *SMA News* **35**, 12 (2003): 1 & 4.

Baylis F and Robert JS. Part-Human Chimeras: Worrying the Facts, Probing the Ethics, *American Journal of Bioethics* **7**, 5 (2007): 41–45.

Beauchamp TL and Childress JF. *Principles of Biomedical Ethics.* New York: Oxford University Press, 2008 (6th edn).

Beecher HK. Ethics and Clinical Research, *New England Journal of Medicine* **274**, 24 (1966): 1354–1360.

Black M. *Models and Metaphors: Studies in Language and Philosophy.* Ithaca and London: Cornell University Press, 1962.

Blackburn-Starza A. British Scientists Go Abroad over Lack of Funding for Human Admixed Embryo Research, *BioNews,* 12 October 2009.

Boon T. A Historical Perspective on Science Engagement, in *Engaging Science: Thoughts, Deeds, Analysis and Action,* ed. Turney J. London: Wellcome Trust, 2007, pp. 8–13.

Bordet S, Bennett J, Knoppers BM and McNagny KM. The Changing Landscape of Human-Animal Chimera Research: A Canadian Regulatory Perspective, *Stem Cell Research* **4**, 1 (2010): 10–16.

Braithwaite J. Regulating Nursing Homes: The Challenge of Regulating Care for Older People in Australia, *British Medical Journal* **323** (2001): 443–446.

Brennan TA, Rothman DJ, Blank L, Blumenthal D, Chimonas SC, Cohen JJ, Goldman J, Kassirer JP, Kimball H, Naughton J and Smelser N. Health Industry Practices That Create Conflicts of Interest: A Policy Proposal for Academic Medical Centers, *Journal of the American Medical Association* **295**, 4 (2006): 429–433.

Briggle A. The Kass Council and the Politicization of Ethics Advice, *Social Studies of Science* **39**, 2 (2009): 309–326.

Brown R. *Social Psychology, the Second Edition.* London: Collier Macmillan, 1986.

Bryder L. *A History of the 'Unfortunate Experiment' at National Women's Hospital.* Auckland: Auckland University Press, 2009.

Bryson B, ed. *Seeing Further: The Story of Science and the Royal Society.* London: HarperPress, 2010.

C

California Institute for Regenerative Medicine, *Regulations Title 17 California Code of Regulations,* Sections 100010–100110, 2006.

California State Senate: Senate Bill Number 1260 (Standards for Egg Retrieval for Stem Cell Research).

Campbell AV. *The Body in Bioethics.* Oxford and New York: Routledge-Cavendish, 2009.

Campbell C. Research on Human Tissue: Religious Perspectives, in *Research Involving Human Biological Materials: Ethical Issues and Policy Guidance,* Vol. II, National Bioethics Advisory Commission, USA, January 2000, pp. C16–20.

Canadian College of Medical Geneticists, *Guidelines for DNA Banking,* 2008.

Canada Health, *Reimbursement of Expenditures under the Assisted Human Reproduction Act,* Public Consultation Document, 2007.

Canadian Institutes of Health Research, Natural Sciences and Engineering Research Council of Canada, Social Sciences and Humanities Research Council of Canada, *Tri-Council Policy Statement: Ethical Conduct for Research Involving Humans,* 1999 (with 2000, 2002, 2005 amendments).

——— *CIHR Best Practices for Protecting Privacy in Health Research,* September 2005.

——— *Updated Guidelines for Human Pluripotent Stem Cell Research,* 27 June 2007.

Canada Interagency Advisory Panel on Research Ethics. *Draft 2nd Edition of the Tri-Council Policy Statement: Ethical Conduct for Research Involving Humans,* 2008.

——— *Revised Draft 2nd Edition of the Tri-Council Policy Statement: Ethical Conduct for Research Involving Humans,* 2009.

——— *What's New in the TCPS?* Available at: http://www.pre.ethics.gc.ca/policy-politique/initiatives/docs/What's%20New%20in%20the%20TCPS.pdf.

Canadian Legislature: *Act Respecting Assisted Human Reproduction and Related Research,* Bill C-6, 2004.

Canadian Medical Association. Sperm Donor Pool Shrivels When Payments Cease, *Canadian Medical Health Journal* **182**, 3 (2010): 233.

Canada Medical Research Council, *Guidelines on Research Involving Human Subjects,* 1987.

Canadian Statutes: *Assisted Human Reproduction Act,* 2004.

Canada Supreme Court: Renvoi fait par le gouvernement du Québec en vertu de la Loi sur les renvois à la Cour d'appel, L.R.Q., ch. R-23, relativement à la constitutionnalité des articles 8 à 19, 40 à 53, 60, 61 et 68 de la Loi sur la procréation assistée, L.C. 2004, ch. 2 (Dans l'affaire du), 2008 QCCA 1167 (CanLII) (appealed, Suprême Court of Canada, *Attorney General of Canada* v. *Attorney General of Quebec,* 2008-08-26, 32750).

Cao C. Making Singapore a Research Hub, *Science* **316**, 5830 (2007): 1423–1424.

Cain DM, Loewenstein G and Moore DA. The Dirt on Coming Clean: Possible Effects of Disclosing Conflicts of Interest, *Journal of Legal Studies* **34**, 1 (2005): 1–24.

Capps B and Campbell A. Why (only some) Compensation for Oocyte Donation for Research Makes Ethical Sense, *Journal of International Biotechnology Law* **4** (2007): 89–102.

Carruthers BG and Halliday TC. Negotiating Globalization: Global Scripts and Intermediation in the Construction of Asian Insolvency Regimes, *Law & Social Inquiry* **31**, 3 (2006): 521–584.

Cartwright SR. *The Report of the Committee of Inquiry into Allegations Concerning the Treatment of Cervical Cancer at National Women's Hospital and into Other Related Matters.* Auckland: Government Printing Office, 1988.

Carver T and Pikalo J. Editors' Introduction, in *Political Language and Metaphor: Interpreting and Changing the World,* eds. Carver T and Pikalo J. London and New York: Routledge, 2008, pp. 1–12.

Chan LT. *The Discourse on Foxes and Ghosts: Ji Yun and Eighteenth-Century Literati Storytelling.* Hong Kong: Chinese University Press, 1998.

Chan WT (comp. and trans.). *A Source Book in Chinese Philosophy.* Princeton, NJ: Princeton University Press, 1963.

Chang AL. On the Harvest Bandwagon, *The Straits Times,* 6 June 1999, p. 36.

Chang AL and Soh N. Ethics Trail Medical Research, *The Straits Times,* 30 March 2001, p. H18.

Chang AL. No to Human Cloning But Yes to Some Stem Cell Work, *The Straits Times,* 3 September 2004, p. 3.

Chang AL and Loo D. Proposed Research Privacy Laws Hailed, *The Straits Times*, 19 June 2006, p. H4.
Chang AL. Speak Up on Science Issues Before It's Too Late, *The Straits Times*, 11 August 2009, p. A2.
Chen HB. Human Face Beast Body, *Lianhe Zaobao* (联合早报), 1 July 2007.
Chen HF. Review Board a Must for Institutions Conducting Human Biomed Research, *The Business Times*, 24 November 2004, p. 1.
——— Human Biomed Research on a Roll Here, *The Business Times*, 3 December 2004, p. 2.
——— No Coercion Allowed in Genetic Testing: BAC, *The Business Times*, 26 November 2005, p. 8.
——— New Guidelines for Research Involving Genetic Modification, *The Business Times*, 19 May 2006, p. 14.
——— Biomedical Research Data Off-limits to 3rd Parties: BAC, *The Business Times*, 15 June 2006, p. 14.
Chinese Medical Doctor Association. Birth of Human-Animal Hybrid Embryo, 3 April 2008. Available at: http://www.cmda.net/info/showinfo.asp?id=5520.
Chia A. Biomedical Hub Will Provide At Least 4,000 Jobs: Philip Yeo, *The Business Times*, 11 May 2002, p. 2.
Chren MM and Landefeld CS. Physicians' Behavior and Their Interactions With Drug Companies: A Controlled Study of Physicians Who Requested Additions to a Hospital Drug Formulary, *Journal of the American Medical Association* **271**, 9 (1994): 684–689.
Chuang PM. Biomed Ethics Guide: Open Approach, *The Business Times*, 13 September 2001, p. 14.
Cleary JD. Impact of Pharmaceutical Sales Representatives on Physician Antibiotic Prescribing, *Journal of Pharmacy Technology* **8** (1992): 27–29.
Collins HM and Evans R. The Third Wave of Science Studies: Studies of Expertise and Experience, in *The Philosophy of Expertise*, eds. Selinger E and Crease RP. New York: Columbia University Press, 2006, pp. 39–110.
Colman A. Stem Cell Research in Singapore, *Cell* **132** (2008): 519–521.
Conner S. Vital Embryo Research Driven Out of Britain: Scientists Abandon Plan to Develop Stem Cells After Funding Dries Up, *The Independent*, 5 October 2009.
Council for International Organizations of Medical Sciences, *International Ethical Guidelines for Biomedical Research Involving Human Subjects*, 1982.
——— *The Declaration of Inuyama: Human Genome Mapping, Genetic Screening and Gene Therapy*, 1990.
——— *International Guidelines for Ethical Review of Epidemiological Studies*, 1991.
——— *International Ethical Guidelines for Biomedical Research Involving Human Subjects*, 1991.
——— *International Ethical Guidelines for Biomedical Research Involving Human Subjects*, 1993.
——— *International Ethical Guidelines for Biomedical Research Involving Human Subjects*, 2002.

——— *International Ethical Guidelines for Epidemiological Studies*, 2009.
Cyranoski D. Korea's Stem-Cell Stars Dogged by Suspicion of Ethical Breach, *Nature* **429**, 6987 (2004): 3.

D

Daley GQ, Ahrlund Shef L, Auerbach JM, Benvenisty N, Charo RA, Chen G, Deng HK, Goldstein LS, Hudson KL, Hyun I, Junn SC, Love J, Lee EH, McLaren A, Mummery CL, Nakatsuji N, Racowsky C, Rooke H, Rossant J, Scholer HR, Solbakk JH, Taylor P, Trounson AO, Weissman IL, Wilmut I, Yu J and Zoloth L. The ISSCR Guidelines for Human Embryonic Stem Cell Research, *Science* **315** (2007): 603–604.
Daley GQ, Rooke HM and Witty N. Global Forum Discusses Stem Cell Research Strategy, *Cell Stem Cell* **2**, 5 (2008): 435–436.
Dana J and Loewenstein G. A Social Science Perspective on Gifts to Physicians from Industry, *Journal of the American Medical Association* **290**, 2 (2003): 252–255.
Danish Council of Ethics, *Man or Mouse? Ethical Aspects of Chimaera Research*, 21 June 2007.
Darwin C. *The Descent of Man and Selection in Relation to Sex*. London: John Murray, 1871.
Daugherty CK, Ratain MJ, Minami H, Banik DM, Vogelzang NJ, Stadler WM and Siegler M. Study of Cohort-Specific Consent and Patient Control in Phase I Cancer Trials, *Journal of Clinical Oncology* **16**, 7 (1998): 2305–2312.
Dodds S and Thomson C. Bioethics and Democracy: Competing Roles of National Bioethics Organisations, *Bioethics* **20**, 6 (2006): 326–338.
Droge P, Tan KP, Qiu SJ and Uttam S. Seeding A Second Green Revolution, *The Straits Times*, 2 January 2010, p. B17.
Dunbar R, Barrett L and Lycett J. *Evolutionary Psychology: A Beginner's Guide*. Oxford: Oneworld, 2007.
Dworkin R. *Taking Rights Seriously*. Cambridge, MA: Harvard University Press, 1977.
Dzur AW and Levin D. The 'Nation's Conscience': Assessing Bioethics Commissions as Public Forums, *Kennedy Institute of Ethics Journal* **14**, 4 (2004): 345–348.
——— The Primacy of the Public: In Support of Bioethics Commissions as Deliberative Forums, *Kennedy Institute of Ethics Journal* **17**, 2 (2007): 133–142.

E

Eiseman E. *The National Bioethics Advisory Commission: Contributing to Public Policy*. Arlington, VA: RAND, 2003.
Elliott JM. Ethical Considerations in Human Stem Cell Research, in *Life Sciences: Law and Ethics*, eds. Kaan T and Liu ET, Singapore: Singapore Academy of Law and Bioethics Advisory Committee, 2006, pp. 54–75.
Emmet D. *Rules, Roles and Relations*. London: Macmillan, 1966.

Evans JH. Between Technocracy and Democratic Legitimation: A Proposed Compromise Position for Common Morality Public Bioethics, *Journal of Medicine and Philosophy* **31** (2006): 213–234.

F

Finkel SE. *Causal Analysis with Panel Data.* Thousand Oaks, CA: Sage, 1985.

Flitner D. *The Politics of Presidential Commissions.* Dobbs Ferry, NY: Transnational Publishers, 1986.

Flory J and Emanuel E. Interventions to Improve Research Participants' Understanding in Informed Consent for Research: A Systematic Review, *Journal of the American Medical Association* **292**, 13 (2004): 1593–1601.

Fost N and Levine RJ. The Dysregulation of Human Subjects Research, *Journal of the American Medical Association* **298**, 18 (2007): 2196–2198.

Fourcade M and Savelsberg JJ. Introduction: Global Processes, National Institutions, Local Bricolage: Shaping Law in an Era of Globalization, *Law & Social Inquiry* **31**, 3 (2006): 513–519.

Fox RC and Swazey JP. *Observing Bioethics.* New York: Oxford University Press, 2008.

Fried C. *Medical Experimentation: Personal Integrity and Social Policy.* New York: American Elsevier, 1974.

Friele MB. Do Committees Ru(i)n the Bio-Political Culture? On the Democratic Legitimacy of Bioethics Committees, *Bioethics* **17**, 4 (2003): 301–318.

G

Galton F. *Inquiries into Human Faculty and its Development.* London: Macmillan, 1883.

———*Inquiries into Human Faculty and its Development.* London: Macmillan (2nd edn.), 1893.

Gereffi G. Paths of Industrialization: An Overview, in *Manufacturing Miracles: Paths of Industrialization in Latin America and East Asia*, eds. Gereffi G and Wyman DL. Princeton, NJ: Princeton University Press, 1990, pp. 3–31.

Glannon W. Phase I Oncology Trials: Why the Therapeutic Misconception Will Not Go Away, *Journal of Medical Ethics* **32**, 5 (2006): 252–255.

Glover J. *Choosing Children: Genes, Disability and Design.* Oxford: Oxford University Press, 2006.

Godoy M, Palca J and Novey B. Key Moments in the Stem Cell Debate, *National Public Radio*, 20 November 2007.

Goh CL. Singapore's Quiet Lobbyists, *The Straits Times*, 28 October 2006, pp. S4–S5.

Goh SY. All Food Here, Including GM Food, Safe to Eat: AVA, *The Straits Times*, 25 March 2008, p. H6.

Gomez J. Freedom of Expression and the Media — Part of a Series of Baseline Studies on Seven Southeast Asian Countries, *Article 19*, December 2005, London.

Gottweis H and Triendl R. South Korean Policy Failure and the Hwang Debacle, *Nature Biotechnology* **24** (2006): 141–143.
Giudice L, Santa E and Pool R. *Assessing the Medical Risks of Human Oocyte Donation for Stem Cell Research*, Washington, DC: National Academies Press, 2007.
Greely HT. Defining Chimeras... and Chimeric Concerns, *American Journal of Bioethics* **3**, 3 (2003): 17–20.
Green RM. For Richer or Poorer? Evaluating the President's Council on Bioethics, *HealthCare Ethics Committee Forum* **18**, 2 (2006): 108–124.
Greene M, Schill K, Takahashi S, Bateman-House A, Beauchamp T, Bok H, Cheney D, Coyle J, Deacon T, Dennett D, Donovan P, Flanagan O, Goldman S, Greely H, Martin L, Miller E, Mueller D, Siegel A, Solter D, Gearhart J, McKhann G and Faden R. Moral Issues of Human–Non-Human Primate Neural Grafting, *Science* **309**, 5733 (2005): 385–386.
Gu BC. A New Era in China's Demographic Dynamics, in *The China Population and Labor Yearbook: The Approaching Lewis Turning Point and Its Policy Implications*, Vol. 1, eds. Cai F and Du Y. Holland: Brill Academic Publishers, 2009, pp. 1–32.

H

Hall D and Ames R. *Thinking from the Han: Self, Truth and Transcendence in Chinese and Western Cultures*. Albany: SUNY Press, 1998.
Halliday TC. Recursivity of Global Normmaking: A Sociolegal Agenda, *Annual Review of Law and Social Sciences* **5** (2009): 263–289.
Harris J. *Enhancing Evolution: The Ethical Case for Making Better People*. Princeton, NJ: Princeton University Press, 2007.
Healy J and Braithwaite J. Designing Safer Health Care Through Responsive Regulation, *Medical Journal of Australia* **184** (2006): S56–S59.
Hedgecoe A. Bioethics and the Reinforcement of Socio-technical Expectations, *Social Studies of Science* **40**, 2 (2010): 163–186.
Hill AB. Medical Ethics and Controlled Trials, *British Medical Journal* **1** (1963): 1043–1049.
Highfield R. Human-Pig Hybrid Embryos Given Go Ahead, *Telegraph*, 1 July 2008.
Ho WC. Governing Cloning: United Nations' Debates and the Institutional Context of Standards, in *Contested Cells: Global Perspectives on the Stem Cell Debates*, eds. Capps B and Campbell A. In press: World Scientific/Imperial College London, 2010.
Ho WC, Capps B and Voo TC. Stem Cell Science and its Public: The Case of Singapore, *East Asian Science, Technology and Society: An International Journal* **39**, 1 (2010): (in press).
Hoffmann DE and Tarzian AJ. The Role and Legal Status of Health Care Ethics Committees in the United States, in *Legal Perspectives in Bioethics*, eds. Iltis AS, Johnson SH and Hinze BA. New York and London: Routledge, 2008, pp. 46–67.
Hogarth S. The Regulation of Nutrigenetic Testing: A Role for Civil Society Organisations? *Health Law Review*, 22 June 2008 (Special Edition).

Holm S and Takala T. High Hopes and Automatic Escalators: A Critique of Some New Arguments in Bioethics, *Journal of Medical Ethics* **33**, 1 (2007): 1–4.

Horng S and Grady C. Misunderstanding in Clinical Research: Distinguishing Therapeutic Misconception, Therapeutic Misestimation and Therapeutic Optimism, *IRB* **25**, 1 (2003): 11–16.

Howard JA. Social Psychology of Identities, *Annual Review of Sociology* **26** (2000): 367–393.

Howe J and Landau I. 'Light Touch' Labour Regulation by State Governments in Australia, *Melbourne University Law Review* **16** (2007).

Human Genome Organisation, *Statement on the Principled Conduct of Genetics Research*, 1996.

——— *Statement on DNA Sampling: Control and Access*, 1998.

——— *Statement on Cloning*, 1999.

——— *Statement on Benefit Sharing*, 2000.

——— *Statement on Gene Therapy and Research*, 2001.

——— *Statement on Human Genomic Databases*, 2002.

——— *Statement on Stem Cells*, 2004.

——— *Statement on Pharmacogenomics: Solidarity, Equity and Governance*, 2007.

I

International Society for Stem Cell Research, *Guidelines for the Conduct of Human Embryonic Stem Cell Research*, 21 December 2006.

——— Ethical Standards for Human-to-Animal Chimera Experiments in Stem Cell Research, *Cell Stem Cell* **1**, 4 (2007): 159–163.

——— *Guidelines for the Clinical Translation of Stem Cells*, 2008.

——— Ethics Report on Interspecies Somatic Cell Nuclear Transfer Research, *Cell Stem Cell* **5**, 1 (2009): 27–30.

Irving DN. What is 'Bioethics'? in *Life and Learning X: Proceedings of the Tenth University Faculty for Life Conference*, ed. Koterski JW. Washington, DC: University Faculty for Life, 2002.

Isasi RM and Annas GJ. Arbitrage, Bioethics and Cloning: The ABCs of Gestating a United Nations Cloning Convention, *Case Western Reserve Journal of International Law* **35** (2003): 397–414.

Isasi RM and Knoppers BM. Beyond the Permissibility of Embryonic and Stem Cell Research: Substantive Requirements and Procedural Safeguards, *Human Reproduction* **21**, 10 (2006): 2474–2481.

Isasi RM and Knoppers BM. Monetary Payments for the Procurement of Oocytes for Stem Cell Research: In Search of Ethical and Political Consistency, *Stem Cell Research* **1** (2007): 37–44.

Isasi RM and Nguyen MT. The Rationale for a Registry of Clinical Trials Involving Human Stem Cell Therapies, *Health Law Review* **16**, 2 (2008): 56–68.

Isasi RM and Knoppers BM. Governing Stem Cell Banks and Registries: Emerging Issues, *Stem Cell Research* **3** (2009): 96–105.

J

Japan Ministry of Education, Culture, Sports, Science and Technology, Japan. Background Discussion on the Purposes and Uses of Human Cloned Embryos. Available at: http://.www.mext.go.jp/b_menu/shingi/gijyutu/gijyutu1/006/gijiroku/06060909/001.htm.

—— Notice on Bioethics and Safety Considerations. Available at: http://www.mext.go.jp/a_menu/kagaku/chousei/unyo/kobo/13/koubo13/05/0501.htm.

Jasanoff S. *Designs on Nature: Science and Democracy in Europe and the United States*. Princeton: Princeton University Press, 2005.

Johnson P. *Intellectual Capital for Communities in the Knowledge Economy*. Singapore: National Library Board, 2008.

Jones D. Moral Psychology: The Depths of Disgust, *Nature* **447**, 7146 (2007): 768–771.

Jones DA. What Does the British Public Think About Human-Animal Hybrid Embryos? *Journal of Medical Ethics* **35**, 3 (2009): 168–170.

Jones RV. *Most Secret War: British Scientific Intelligence 1939–1945*. London: Hamish Hamilton, 1978.

Jonsen AR. *The Birth of Bioethics*. New York: Oxford University Press, 1998.

K

Kass LR. The Wisdom of Repugnance, *New Republic* **216**, 22 (1997): 17–26.

Keith A. *A New Theory of Human Evolution*. London: Watts, 1948.

Kelley M, Rubens CE and the GAPPS Review Group. Global Report on Preterm Birth and Stillbirth (6 of 7): Ethical Considerations, *BioMed Central Pregnancy and Childbirth* **10**, Suppl. 1, S6 (2010): 1–19.

Kesava S. As Research Grants Grow, Watch Out for Fraud, *The Straits Times*, 23 July 2008, pp. 1–2.

Khalik S. Setting Standards for Ethics of Life-sciences Research, *The Straits Times*, 9 December 2000, p. H6.

King NMP. Defining and Describing Benefit Appropriately in Clinical Trials, *Journal of Law and Medical Ethics* **28**, 4 (2000): 332–343.

Kitzinger J. The Role of Media in Public Engagement, in *Engaging Science: Thoughts, Deeds, Analysis and Action*, ed. Turney J. London: Wellcome Trust, 2007, pp. 45–49.

Knight Ridder Tribune. Senator Orrin Hatch Aborts Views to Support Stem Cell Research, 17 February 2003.

Knoppers BM and Le Bris S. Genetic Choices: A Paradigm for Prospective International Ethics? *Politics and Life Sciences* **13**, 2 (1994): 228–230.

Knoppers BM. Reflections: The Challenges of Biotechnology and Public Policy, *McGill Law Journal* **45** (2000): 559–566.

Knoppers BM and Chadwick R. Human Genetics Research: Emerging Trends in Ethics, *Nature Reviews: Genetics* **6** (2005): 75–79.

Knoppers BM, Richardson G, Lee EH, Kure J, de Wert G, Ameisen JC, Lötjönen L, Breen K, Isasi R, Mauron A, Murray T and Wahlström J (for the ISCF Ethics Working Party). Ethics Issues in Stem Cell Research, *Science* **312** (2006): 366–367.

Knoppers BM, Revel M, Richardson G, Kure J, Lötjönen S, Isasi R, Mauron A, Wahlstrom J, Rager B and Peng PL (for the ISCF Ethics Working Party). Letter to the Editor: Oocyte Donation for Stem Cell Research, *Science* **316** (2007): 368–370.

Knoppers BM. Challenges to Ethics Review in Health Research, *Health Law Review* **17**, 2&3 (2009): 47.

Knoppers BM. Genomics and Policymaking: From Static Models to Complex Systems? *Human Genetics* **125**, 4 (2009): 375–379, p. 375.

Koh T. The 10 Values that Undergird East Asian Strength and Success, *International Herald Tribune,* 11–12 December 1993, p. 6.

L

Lancet Editors. Retraction — Ileal-lymphoid-nodular hyperplasia, non-specific colitis, and pervasive developmental disorder in children, *Lancet,* **375**, 9713 (2010): 445.

Lansford JE, Deater-Deckard K, Dodge KA, Bates JE and Pettit GS. Ethnic Differences in the Link Between Physical Discipline and Later Adolescent Externalizing Behaviours, *Journal of Child Psychology and Psychiatry* **45**, 4 (2004): 801–812.

Latour B. Scientific Objects and Legal Objectivity, in *Law, Anthropology, and the Constitution of the Social: Making Persons and Things,* eds. Pottage A and Mundy M. Cambridge: Cambridge University Press, 2004, pp. 73–114.

Levine RJ. New International Ethical Guidelines for Research Involving Human Subjects. *Annals of Internal Medicine* **119**, 4 (1993): 339–341.

Levine AD. Identifying Under- and Overperforming Countries in Research Related to Human Embryonic Stem Cells, *Cell Stem Cell* **2,** 6 (2008): 521–524.

Levy D. *Love and Sex with Robots: The Evolution of Human-Robot Relations.* New York: HarperCollins, 2007.

Li C. *The Tao Encounters the West.* Albany: SUNY Press, 1999, p. 94.

Lim CP. Betting on Biomedical Science: The Nation's Economy Has Evolved Rapidly in Just a Few Decades from Labor-intensive Manufacturing to High-tech Production and Now to Corporate Management and World-class Research, *Issues in Science and Technology,* 22 March 2010.

Lim LPL and Gregory MJ. Singapore's Biomedical Science Sector Development Strategy: Is it Sustainable? *Journal of Commercial Biotechnology* **10**, 4 (2004): 352–362.

Lin M. *Certainty as Social Metaphor: The Social and Historical Production of Certainty in China and the West.* Westport, CT, and London: Greenwood Press, 2001.

Lo B, Wolf LE and Berkeley A. Conflict-of-Interest Policies for Investigators in Clinical Trials, *New England Journal of Medicine* **343**, 22 (2000): 1616–1620.

Lo B, Chou V, Cedars MI, Gates E, Taylor RN, Wagner RM, Wolf L and Yamamoto KR. Consent from Donors for Embryo and Stem Cell Research, *Science* **301** (2003): 921.

Lo B and Parham L. Ethical Issues in Stem Cell Research, *Endocrine Reviews* **30**, 3 (2009): 204–213.

Lo B, Parham L, Broom C, Cedars M, Gates E, Giudice L, Halme DG, Hershon W, Kriegstein A, Kwok PY, Oberman M, Roberts C and Wagner R. Importing Human Pluripotent Stem Cell Lines Derived at Another Institution: Tailoring Review to Ethical Concerns, *Cell Stem Cell* **4** (2009): 115–123.

Lo B. *Ethical Issues in Clinical Research: A Practical Guide*. Philadelphia, PA: Lippincott Williams & Wilkins, 2010.

Lomax GP, Hall ZH and Lo B. Responsible Oversight of Human Stem Cell Research: The California Institute for Regenerative Medicine's Medical and Ethical Standards, *Public Library of Science (PLoS) Medicine* **4**, 5 (2007): e114.

Loo D. More Drug Companies Conducting Trials in S'pore, *The Straits Times*, 9 September 2006, p. 2.

——— Research Drive On The Right Track, Says Tharman, *The Straits Times*, 16 February 2007, p. 3.

Low E. Set Up Body to Regulate Stem Cell Research: Panel, *The Business Times*, 8 January 2002, p. 7.

M

Macer D and Chin CO. Bioethics Education Among Singapore High School Science Teachers, *Eubios Journal of Asian and International Bioethics* **9** (1999): 138–144.

Maclaren K. Emotional Metamorphoses: The Role of Others in Becoming a Subject, in *Embodiment and Agency*, eds. Campbell S, Meynell L and Sherwin S. University Park, PA: Pennsylvania State University, 2009, pp. 25–45.

Maden C. Human/Animal Hybrid Embryos are 'Easy' to Make, *BioNews*, 23 June 2008.

Manning J. *The Cartwright Papers: Essays on the Cervical Cancer Inquiry of 1987–88*, Wellington: Bridget Williams Books, 2009.

Matterson C. Engaging Science: Creative Enterprise or Controlled Endeavour? in *Engaging Science: Thoughts, Deeds, Analysis and Action*, ed. Turney J. London: Wellcome Trust, 2007, pp. 8–13.

Mathews DJH, Donovan P, Harris J, Lovell-Badge R, Savulescu J and Faden R. Integrity in International Stem Cell Research Collaborations, *Science* **313** (2006): 921–922.

May L. *Sharing Responsibility*. Chicago: University of Chicago Press, 1992.

——— *The Socially Responsive Self: Social Theory and Professional Ethics*. Chicago: University of Chicago Press, 1996.

McCoy P. Govt Biomedical Watchdog Body May Be Set Up, *The Straits Times*, 18 November 2001, p. 5.

McLaren A. Free-Range Eggs? *Science* **316**, 5823 (2007): 339.

de Melo-Martín I and Ho A. Beyond Informed Consent: The Therapeutic Misconception and Trust, *Journal of Medical Ethics* **34**, 3 (2008): 202–205.

Merz JF and Fischhoff B. Informed Consent Does Not Mean Rational Consent: Cognitive Limitations on Decision-Making, *Journal of Legal Medicine* **11**, 3 (1990): 321–350.

Meslin EM and Johnson S. National Bioethics Commissions and Research Ethics, in *The Oxford Textbook of Clinical Research Ethics*, eds. Emanuel EJ, Grady C, Crouch RA, Lie R, Miller F and Wendler D. New York: Oxford University Press, 2008, pp. 187–197.

Milbrath LW and Goel ML. *Political Participation*. Chicago: Rand McNally, 1977.

Miller FG and Joffe S. Evaluating the Therapeutic Misconception, *Kennedy Institute of Ethics Journal* **16**, 4 (2006): 353–366.

Miller FG. Consent to Clinical Research, in *The Ethics of Consent: Theory and Practice*, eds. Miller FG and Wertheimer A. Oxford: Oxford University Press, 2010, pp. 375–404.

Miller FH. Trusting Doctors: Tricky Business When It Comes to Clinical Research, *Boston University School of Law Working Paper Series*, Public Law & Legal Theory: Working paper no. 01-09.

Moravec H. Rise of the Robots — The Future of Artificial Intelligence, *Scientific American* **18** (February 2008): 12–19.

Morin K, Rakatansky H, Riddick Jr. FA, Morse LJ, O'Bannon III JM, Goldrich MS, Ray P, Weiss M, Sade RM and Spillman MA. Managing Conflicts of Interest in the Conduct of Clinical Trials, *Journal of the American Medical Association* **287**, 1 (2002): 78–84.

Morreim EH. Medical Research Litigation and Malpractice Tort Doctrines: Courts on a Learning Curve. *Houston Journal of Health Law and Policy* **4**, 1 (2003): 1–86.

———— The Clinical Investigator as Fiduciary: Discarding a Misguided Idea, *Journal of Law and Medical Ethics* **33**, 3 (2005): 586–598.

Mottier V. Metaphors, Mini-narratives and Foucaldian Discourse Theory, in *Political Language and Metaphor: Interpreting and Changing the World*, eds. Carver T and Pikalo J. London and New York: Routledge, 2008, pp. 182–194.

Mulkay M. *The Embryo Research Debate: Science and the Politics of Reproduction*. Cambridge: Cambridge University Press, 1997.

Murray TH. Genetic Exceptionalism and "Future Diaries": Is Genetic Information Different From Other Medical Information? in *Genetic Secrets: Protecting Privacy and Confidentiality in the Genetic Era*, ed. Rothstein M. New Haven: Yale University Press, 1997, pp. 60–73.

N

Nature Neuroscience. Editorial, Primate Rights? **10**, 6 (2007): 669.

Nisbet M. Who's Getting It Right and Who's Getting It Wrong in the Debate about Science Literacy? Opinions Clash Over the Best Way to Bolster Public Support for Science, *Science and the Media*, 9 June 2003.

Nong Bo Web. Can You Accept Human-Animal Combinations? 1 September 2008. Available at: http://biology.aweb.com.cn/news/2008/1/9/1112193.shtml.

Normile D. Can Money Turn Singapore Into a Biotech Juggernaut? *Science* **297**, 5586 (2002): 1470–1473.

Normile D. An Asian Tiger's Bold Experiment, *Science* **316**, 5821 (2007): 38–41.

Nozick R. *Anarchy, State and Utopia*. New York: Basic Books, 1974.

Nuremberg Military Tribunal. *Trials of War Criminals before the Nuremberg Military Tribunals under Control Council Law*, No. 10, Vol. 2. Washington, DC: US Government Printing Office, 1949, pp. 181–182.

Nuyen AT. Stem Cell Research and Interspecies Fusion: Some Philosophical Issues, 2007. Available at: www.bioethics-singapore.org.

——— Moral Obligations and Moral Motivation in Confucian Role-Based Ethics, *Dao* **8** (2009): 1–11.

O

O'Neill O. *Autonomy and Trust in Bioethics*. Cambridge: Cambridge University Press, 2002.

Ohorodnyk P, Eisenhauer EA and Booth CM. Clinical Benefit in Oncology Trials: Is This a Patient-Centred or Tumour-Centred End-point? *European Journal of Cancer* **45**, 13 (2009): 224–225.

Organization for Economic Co-operation and Development, *OECD Guidelines for Quality Assurance in Molecular Genetic Testing*, 2007.

——— *OECD Guidelines for Human Biobanks and Genetic Research Databases*, 2009.

P

Pappworth MH. *Human Guinea Pigs: Experimentation on Man*. London: Routledge & Kegan Paul, 1967.

Pearce RD. *The Turning Point in Africa: British Colonial Policy, 1938–48*. London: Cass, 1982.

Peay MY and Peay ER. The Role of Commercial Sources in the Adoption of a New Drug, *Social Science and Medicine* **26**, 12 (1988): 1183–1189.

Pew Research Center. *Most Want Middle Ground on Abortion: Pragmatic Americans Liberal and Conservative on Social Issues*, 3 August 2006.

Pielke RA Jr. *The Honest Broker: Making Sense of Science in Policy and Politics*. Cambridge: Cambridge University Press, 2007.

R

Rachels J. *Created from Animals: The Moral Implications of Darwinism*. Oxford: Oxford University Press, 1990.

Rawls J. *A Theory of Justice*. Oxford: Clarendon Press, 1972.

Redfern M, Keeling JW and Powell E, *The Royal Liverpool Children's Inquiry Report*, UK: HM Statwnevy Office, 2001.

Reuters. Medical Journal Retracts Autism Paper 12 Years On, 2 February 2010.

Riles A. *The Network Inside Out*. Michigan: University of Michigan Press, 2001.

——— *Collateral Knowledge: Legal Reasoning in the Global Financial Markets*. In press, 2010.

Rivard G. Article 11: Non-Discrimination and Non-Stigmatization, in *The UNESCO Universal Declaration on Bioethics and Human Rights: Backgrounds, Principles and Applications*, eds. ten Have HAMJ and Jean MS. Paris: UNESCO Publishing, 2009, pp. 187–198.

Rugg-Gunn PJ, Ogbogu U, Rossant J and Caulfield T. The Challenge of Regulating Rapidly Changing Science: Stem Cell Legislation in Canada, *Cell Stem Cell* **3**, 4 (2009): 285–288.

Rynning E. Legal Tools and Strategies for the Regulation of Chimbrids, in *CHIMBRIDS — Chimeras and Hybrids in Comparative European and International Research*, eds. Taupitz J and Weschka M. Heidelberg: Springer, 2009, pp. 79–87.

S

Santos BS. *Law and Globalization from Below: Towards a Cosmopolitan Legality*. Cambridge: Cambridge University Press, 2005.

Sasaki E, Suemizu H, Shimada A, Hanazawa K, Oiwa R, Kamioka M, Tomioka I, Sotomaru Y, Hirakawa R, Eto T, Shiozawa S, Maeda T, Ito M, Ito R, Kito C, Yagihashi C, Kawai K, Miyoshi H, Tanioka Y, Tamaoki N, Habu S, Okano H and Nomura T. Generation of Transgenic Non-human Primates with Germline Transmission, *Nature* **459**, 7246 (2009): 523–528.

Seah L. When I Grow up, I Don't Want To Be A Scientist. Where Have All The Young Researchers Gone? *The Straits Times*, 2 November 1995, pp. L1–L2.

Searle JR. Minds, Brains, and Programs, *Behavioral and Brain Sciences* **3**, 3 (1980): 417–457.

Shah ZA. *The Astronomical Calculations and Ramadan: A Fiqhi Discourse*. Washington and London: International Institute of Islamic Thought, 2009.

Shaikh N. Lack of Funding Threatens Progress in Admixed Embryo Research, *BioNews*, 19 January 2009.

Shepherd E. *Human Fertilisation and Embryology Bill [HL], HL Bill 6, 2007–08*. London: House of Lords Library Notes, 14 November 2007.

Shorvon S. The Prosecution of Research — Experience from Singapore, *The Lancet* **369**, 9576 (2007): 1835–1837.

Singapore Bioethics Advisory Committee, *Human Stem Cell Research Consultation Paper*, 8 November 2001.

——— *Ethical, Legal and Social Issues in Human Stem Cell Research, Reproductive and Therapeutic Cloning*, June 2002.

——— *Human Tissue Research*, November 2002.

——— *Advancing the Framework on Ethics Governance for Human Research: A Consultation Paper*, 16 September 2003.

——— *Research Involving Human Subjects: Guidelines for IRBs*, November 2004.

——— *Genetic Testing and Genetic Research*, November 2005.

——— *Personal Information in Biomedical Research*, May 2007.

——— *Donation of Human Eggs for Research: A Consultation Paper*, 7 November 2007.

——— *Human-Animal Combinations for Biomedical Research: A Consultation Paper*, 8 January 2008.

——— *Donation of Human Eggs for Research*, November 2008.

Singapore Economic Strategies Committee, *High-Skilled People, Innovative Economy, Distinctive Global City*, February 2010.

Singapore Ministry of Health, *Singapore Guideline for Good Clinical Practice*, 1998, revised 1999.

——— *Directive 1A/2006: BAC Recommendations for Biomedical Research*, 18 January 2006.

——— *Directives for Private Healthcare Institutions Providing Assisted Reproduction Services: Regulation 4 of the Private Hospitals and Medical Clinics Regulations* (Cap. 248, Reg. 1), September 2001; Revised March 2006.

Singapore National Medical Ethics Committee, *Ethical Guidelines on Research Involving Human Subjects*, August 1997.

——— *Ethical Guidelines for Gene Technology*, February 2001.

Singapore Regulations: *Medicines (Clinical Trials) (Amendment) Regulations*, 2000 Revised Edition.

Singapore Statutes: *Constitution of the Republic of Singapore*.

——— *Administration of Muslim Law Act* (Cap. 3), Revised 2009.

——— *Maintenance of Religious Harmony Act* (Cap. 167A), Revised 2001.

——— *Medical (Therapy, Education and Research) Act* (Cap. 175), Revised 1985 (Amended 2010).

——— *Medicines Act* (Cap. 176), Revised 1985.

——— *Human Cloning and Other Prohibited Practices Act* (Cap. 131B), Revised 2005.

——— *National Registry of Diseases Act* (Cap. 201B), Revised 2008.

Singer P. Moral Experts, in *The Philosophy of Expertise*, eds. Selinger E and Crease RP. New York: Columbia University Press, 2006, pp. 187–189.

Skene L. Undertaking Research in Other Countries: National Ethico-Legal Barometers and International Ethical Consensus Statements, *Public Library of Science (PLoS) Medicine* **4** (2007): 0243–0247.

Skidelsky R. The Price of Clarity, *The Straits Times*, 24 May 2010, p. A24.

Smiley M. *Moral Responsibility and the Boundaries of Community: Power and Accountability from a Pragmatic Point of View*. Chicago: University of Chicago Press, 1992.

Soh N and Chang AL. Panel Sought Various Views for Guidelines, *The Straits Times*, 22 June 2002, p. H19.

Sreenivasan G. Does Informed Consent to Research Require Comprehension? *Lancet* **362**, 9400 (2003): 2016–2018.

Steinbrook R. The Gelsinger Case, in *The Oxford Textbook of Clinical Research Ethics*, eds. Emanuel EJ *et al.* New York: Oxford University Press, 2008, pp. 110–120.

Stenvoll D. Slippery Slopes in Political Discourse, in *Political Language and Metaphor: Interpreting and Changing the World*, eds. Carver T and Pikalo J. London and New York: Routledge, 2008, pp. 28–40.

Stiller CA. Centralised Treatment, Entry to Trials and Survival, *British Journal of Cancer* **70**, 2 (1994): 352–362.

Straits Times. Few People Here Aware of GM Products, 23 November 2000, p. H13.

Strathern M. Accountability... and Ethnography, in *Audit Cultures: Anthropological Studies in Accountability, Ethics and the Academy*, ed. Strathern M. London: Routledge, 2000, pp. 279–304.

Sunstein CR. Health-Health Tradeoffs, in *Risk and Reason*. Cambridge: Cambridge University Press, 2002, pp. 133–152.

T

Takahashi K, Tanabe K, Ohnuki M, Narita M, Ichisaka T, Tomoda K and Yamanaka S. Induction of Pluripotent Stem Cells from Adult Human Fibroblasts by Defined Factors, *Cell* **131**, 5 (2007): 1–12.

Tan T. Forum on Ethics of Mixing Human, Animal Genes, *The Straits Times*, 14 August 2008.

Taverne D. Let's Be Sensible About Public Participation, *Nature* **432**, 7015 (2004): 271.

Taylor C. *Sources of the Self*. Cambridge, MA: Harvard University Press, 1989.

Triggle N. Lancet Accepts MMR Study 'False', *BBC*, 2 February 2010.

Tu WM. *Centrality and Commonality: An Essay on Confucian Religiousness*. Albany: SUNY Press, 1989.

U

UK Academy of Medical Sciences, *Inter-species Embryos: A Report by the Academy of Medical Sciences*, June 2007.

UK Committee on Human Fertilisation and Embryology, *Report of the Committee of Enquiry Into Human Fertilisation and Embryology*, Warnock M (Chair), 1984.

UK Committee on Standards in Public Life, *The First Seven Reports: A Review of Progress*, September 2001.

UK Department of Health, *Stem Cell Research: Medical Progress with Responsibility*, June 2000.

——— *Review of the Human Fertilisation and Embryology Act: Proposals for Revised Legislation (Including Establishment of the Regulatory Authority for Tissue and Embryos)*, December 2006.

UK House of Commons Science and Technology Select Committee, *Reproductive Technologies and the Law*, 2005.

UK House of Lords Select Committee, *Stem Cell Research*, 2002.

UK Human Fertilisation and Embryology Authority, *Hybrids and Chimeras: A Report on the Findings of the Consultation*, October 2007.

UK Medical Research Council, Press Release, *Women Undergoing IVF to Donate Eggs for Stem Cell Research in Return for Reduced Treatment Costs*, 13 September 2007.

UK Newcastle University, *Press Release: Hybrid Embryos Statement*, 1 April 2008.
UK Statutes: Human Fertilisation and Embryology Act, 2008 (and Explanatory Notes).
United Nations (League of Nations). *Geneva Declaration of the Rights of the Child*, GA res. 1386 (XIV), 14 UN GAOR Supp. (No. 16), p. 19, UN Doc A/4354, 1924.
——— *International Bill of Human Rights*, GA A/RES/3/217, 10 December 1948.
——— *Universal Declaration of Human Rights*, GA res. 217A (III), UN Doc A/810, p. 71 (1948).
——— *Convention on the Rights of the Child*, GA A/RES/44/25, 12 December 1989.
——— *Declaration on Human Cloning*, GA Res. 59/280, UN GAOR, 59th Sess., UN Doc. A/RES/59/280, 2005.
——— *Declaration on the Rights of Indigenous Peoples*, GA A/RES/61/295, 2 October 2007.
United Nations Educational, Scientific and Cultural Organization, *Universal Declaration on the Human Genome and Human Rights*, 1997.
——— *International Declaration on Human Genetic Data*, 2003.
——— *Universal Declaration on Bioethics and Human Rights*, 2005.
——— International Bioethics Committee, *Report of IBC on Human Cloning and International Governance*, June 2009.
——— *Conclusions of the 16th Session of the Intergovernmental Bioethics Committee*, 10 July 2009.
United Nations University — Institute of Advanced Studies, *Is Human Reproductive Cloning Inevitable: Future Options for UN Governance*, 2007.
USA Congress: *National Research Act*, 12 July 1974, Public Law 93-348.
——— Balanced Budget Down Payment Act, 26 January 1996, Public Law 104-199.
USA Department of Health Education and Welfare, Ethics Advisory Board, *Report and Conclusions: HEW Support of Research Involving Human In Vitro Fertilization and Embryo Transfer*, 4 May 1979.
USA Department of Health and Human Services, Office of Human Research Protection, *Financial Relationships in Interests in Research Involving Human Subjects: Guidance for Human Subject Protection*, 5 May 2004.
USA Food and Drug Administration. FDA Warns Regenerative Sciences About Unlicensed Drug, *FDA News*, 22 August 2008.
USA Government: *Code of Federal Regulations*, Title 45, Part 46, Subparts A–D (Protection of Human Subjects).
——— Establishing the Presidential Commission for the Study of Bioethical Issues, *Federal Register* **74**, 228 (30 November 2009).
USA National Academy of Sciences, *Guidelines for Human Embryonic Stem Cell Research*, 2005 (amended 2007, 2008 and 2010).
USA National Commission for the Protection of Human Subjects of Biomedical and Behavioral Research, *The Belmont Report: Ethical Principles and Guidelines for the Protection of Human Subjects of Research*, 18 April 1979.
USA National Bioethics Advisory Commission, *Research Involving Human Biological Materials: Ethical Issues and Policy Guidance*, Vol. I, August 1999.
——— *Ethical Issues in Human Stem Cell Research*, September 1999.
——— *Research Involving Human Biological Materials: Ethical Issues and Policy Guidance*, Vol. II, January 2000.

USA National Institutes of Health. *Report of the Human Embryo Research Panel.* Bethesda, MD: National Institutes of Health, 1994.

USA National Research Council and Institute of Medicine. *Guidelines for Human Embryonic Stem Cell Research.* Washington, DC: National Academies Press, 2005.

USA President's Commission for the Study of Ethical Problems in Medicine and Biomedical and Behavioral Research, *Securing Access to Health Care: The Ethical Implications of Differences in the Availability of Health Services,* March 1983.

V

Van Epps HL. Singapore's Multibillion Dollar Gamble. *Journal of Experimental Medicine* **203,** 5 (2006): 1139–1142.

Vaughn V. Biomed Sector To Grow By Up To 10%, *The Straits Times,* 18 March 2010, p. B18.

Vestal C. Embryonic Stem Cell Research Divides States, *Stateline.org,* 21 June 2007.

Vist GE, Hagen KB, Devereaux PJ, Bryant D, Kristoffersen DT and Oxman AD. Systematic Review to Determine Whether Participation in a Trial Influences Outcome, *British Medical Journal* **330**, 7501 (2005): 1175.

Voo TC. Therapeutic Misconception, in *International Encyclopaedia of Applied Ethics,* eds. Chadwick R, Callahan D and Singer P. London: Academic Press, forthcoming in 2010.

W

de Waal FBM. *Good Natured: The Origin of Right and Wrong in Humans and Other Animals.* Cambridge, Massachusetts: Harvard University Press, 1996.

——— *Primates and Philosophers: How Morality Evolved.* Princeton, New Jersey: Princeton University Press, 2006.

Waldby C. Singapore Biopolis: Bare Life in the City-State, *East Asian Science, Technology and Society: An International Journal* **3** (2009): 367–383.

Wardrop M. Human-Animal Clone Research Halted Amid Funding Drought: Britain's Research into the Creation of Human-Animal Clones Has Ground to a Halt Due to a Lack of Funding, *Telegraph,* 13 January 2009.

Watts G, ed. *Hype, Hope and Hybrids: Science, Policy and Media Perspectives of the Human Fertilisation and Embryology Bill,* UK: The Academy of Medical Sciences, Medical Research Council, Science Media Centre and Wellcome Trust, 2009.

Wazana A. Physicians and the Pharmaceutical Industry: Is a Gift Ever Just a Gift? *Journal of the American Medical Association* **283**, 3 (2000): 373–380.

Weisbard AJ. The Role of Philosophers in the Public Policy Process: A View from the President's Commission, *Ethics* **97** (1987): 776–785.

Wertz DC and Knoppers BM. The HUGO Ethics Committee: Six Innovative Statements, *GE^3LS* **2**, 1 (2003): 1 & 4.

Wheat K and Matthews K. World Human Cloning Policies, *Innovative e-Learning in Reparative Medicine* (NOVAe-MED).

Whittall H. A Closer Look at the Nuffield Council on Bioethics, *Clinical Ethics* **3** (2008): 199–204.

Williams P. Rights and the Alleged Rights of the Innocents to Be Killed, *Ethics* **87** (1977): 384–394.

Wilsdon J and Willis R. *See-through Science: Why Public Engagement Needs to Move Upstream*. London: Demos, 2004.

Witty N. Strategic Planning: Progress and Potential, *Cell Stem Cell* **1**, 4 (2007): 383–386.

Woo KT. Conducting Clinical Trials in Singapore, *Singapore Medical Journal* **40**, 4 (1999): 310–313.

World Health Organization, *Control of Hereditary Diseases — Report of a WHO Scientific Group*, 1996.

——— *Statement of the WHO Expert Advisory Group on Ethical Issues in Medical Genetics*, 1997.

——— *Proposed International Guidelines on Ethical Issues in Medical Genetics and Genetic Services*, 1998.

——— *Ethical, Scientific and Social Implications of Cloning in Human Health*, 1998.

——— *Cloning in Human Health: Report by the Secretariat (A52/12)*, 1999.

——— *Genomics and World Health: Report of the Advisory Committee on Health Research*, 2002.

——— *Genetic Databases: Assessing the Benefits and the Impact on Human and Patient Rights*, 2003.

——— *Control of Genetic Diseases: Report by the Secretariat (EB116/3)*, 2005.

——— *Genetics, Genomics and the Patenting of DNA: Review of Potential Implications for Health in Developing Countries*, 2005.

World Medical Association, *Declaration of Geneva*, 1948.

——— *Declaration of Helsinki: Recommendations Guiding Doctors in Clinical Research*, 1 June 1967.

——— *Declaration of Helsinki—Ethical Principles for Medical Research Involving Human Subjects*, 22 October 2008 (as amended).

Wright R. *The Moral Animal: Evolutionary Psychology and Everyday Life*. New York: Pantheon Books, 1994.

Wynne B. Public Engagement as a Means of Restoring Public Trust in Science — Hitting the Notes, but Missing the Music? *Community Genetics* **9**, 3 (2006): 211–220.

X

Xie YY. Bioethics Advisory Committee Consults the Public: Can You Accept Human-Animal Combinations, *Lianhe Zaobao* (联合早报), 9 January 2008.

Xu XF and Li TB. Human-Animal Strange Research, Raises All Kinds of Issues: Meat with Human Cells, Do You Dare Eat?, *Xinmin Wanbao* (新民晚报), 8 January 2008.

Xu XY. Human-Animal Combinations? Bioethics Advisory Committee Seeks Public Opinion, *Lianhe Wanbao* (联合晚报), 8 January 2008.

Y

Yanow D. Cognition Meets Action: Metaphors as Models of and Models for, in *Political Language and Metaphor: Interpreting and Changing the World*, eds. Carver T and Pikalo J. London and New York: Routledge, 2008, pp. 225–238.

Yeo G. Don't Let Dogma Block Potential of Stem-Cell Study, Speech at the 35th annual dinner of the Institution of Engineers Singapore, 17 September 2001; reported in *The Straits Times*, 19 September 2001, p. 18.

Yu J, Vodyanik MA, Smuga-Otto K, Antosiewicz-Bourget J, Frane JL, Tian S, Nie J, Jonsdottir GA, Ruotti V, Steward R, Slukvin II and Thomson JA. Induced Pluripotent Stem Cell Lines Derived from Human Somatic Cells, *Science* **318**, 5858 (2007): 1917–1920.

Z

Zakaria F. Culture Is Destiny: A Conversation with Lee Kuan Yew, *Foreign Affairs* **73** (1994): 109–129.

Index